IMMUNE RECOGNITION

M.J.Owen

Imperial Cancer Research Fund Laboratories,
St Bartholomew's Hospital, Dominion House,
Bartholomew Close, London EC1A 7BE, UK

J.R.Lamb

MRC Tuberculosis and Related Infections Unit,
Royal Postgraduate Medical School,
Hammersmith Hospital, DuCane Road, London W12 0HS, UK

IRL PRESS
OXFORD · WASHINGTON DC

Published by:
IRL Press Limited
PO Box 1,
Eynsham,
Oxford OX8 1JJ,
UK

1604817

©1988 IRL Press Limited

British Library Cataloguing in Publication Data

Owen, M.J.
    Immune recognition.
    1. Man. Immune reactions
    I. Title   II. Lamb, J.R.   III. Series
    616.07'95

    ISBN 1-85221-062-1

Typeset by Infotype and printed by Information Printing Ltd, Oxford, England.

Preface

Immune recognition is central to our understanding of the immune system. The ability to specifically recognize molecules on pathogens is a fundamental feature of the adaptive immune response, which is mediated by the antigen receptors on T and B lymphocytes. B lymphocytes use antibody as their antigen receptors and antibodies have been recognized for many years. Their structures have been elucidated in the last 20 years and the antibody genes are well characterized. It is only in the past 6 years that the T cell antigen receptor has been identified and the genes sequenced. The two receptor types show remarkable similarities in the way that enormous numbers of different receptors are generated from a limited number of germline genes. The role of the major histocompatibility complex (MHC) molecules in T cell mediated immune recognition has been appreciated for some time, but it is only now, with the appearance of the first detailed structures of MHC molecules, that we can start to appreciate the molecular interactions which take place at the cell surface when T lymphocytes recognize antigen on the surface of another cell. Many elements of the interactions still need to be determined, and even the functions of some of the molecules involved in recognition (such and the γ and δ T cell receptor chains) still need explanation. Nevertheless, it appears that the major sets of molecules involved in immune recognition have now been identified. It is therefore particularly timely to review in this book, what is known of these molecules and of immune recognition at the molecular level. We may look forward in the next few years to further details of how expression of the molecules is controlled at the cell surface during lymphocyte ontogeny and how recognition is linked to lymphocyte activation or tolerance.

David Male

Contents

3. The T cell antigen receptor

4. T cell recognition of antigen and MHC gene products

Abbreviations

AFC	antibody-forming cell
A-MuLV	Abelson murine leukaemia virus
APC	antigen-presenting cells
DAG	diacylglycerol
GA	glutamic acid – alanine copolymer
HEL	hen egg lysozyme
IFNγ	Interferon-γ
Ig	immunoglobulin
IL	interleukin
IP$_3$	inositol triphosphate
Ir	immune response
LFA	leukocyte function associated
Mab	monoclonal antibody
m.Cytc	moth cytochrome c
MHC	major histocompatibility complex
mIg	membrane Ig
MYO	myoglobin
NK	natural killer
NP	nucleoprotein
OVA	ovalbumin
p.Cytc	pigeon cytochrome c
PInsP$_2$	phosphatidylinositol 4,5-biphosphate
PKC	protein kinase C
PLC	phospholipase C
λrepr.	λ cI repressor
sIg	secreted Ig
SWM	sperm whale myoglobin
T$_C$	cytotoxic T cell
TCR	T cell antigen receptor
T$_H$	helper T cell
Ti	the idiotype$^+$ (antigen-binding) portion of the TCR
T$_S$	suppressor T cell

Amino acids

A	alanine	D	aspartic acid
C	cysteine	E	glutamic acid

F	phenylalanine	P	proline
G	glycine	Q	glutamine
H	histidine	R	arginine
I	isoleucine	S	serine
K	lysine	T	threonine
L	leucine	V	valine
M	methionine	W	tryptophan
N	asparagine	Y	tyrosine

1

General principles of recognition

1. Immune recognition

The immune system in mammals has evolved to recognize and destroy the large variety of potential pathogens which an individual may encounter. Recognition is mediated by T and B lymphocytes. B cells can recognize antigens in their native conformation either free in solution, on membranes, or on the surface of cells, using surface immunoglobulin (Ig) as their specific antigen receptor. The T cell antigen receptor (TCR) is structurally different from antibody, it is generated by different sets of genes and most T cells can only recognize antigen on the surface of other cells. To be precise, T cells usually recognize processed or degraded antigen, and only when it is physically associated with molecules encoded by the major histocompatibility complex (MHC). In effect, MHC molecules act as a guidance system for T cells, allowing them to recognize antigens from within cells. These three groups of molecules, antibody, the TCR and MHC molecules, control the process of immune recognition.

To protect an individual, the lymphocytes must be capable of recognizing any potential pathogen. Since there is a strong selective pressure favouring micro-organisms which can avoid immune recognition (e.g. by mutation of their antigens), the lymphocyte populations as a whole have antigen receptors which can specifically recognize virtually any biological molecule. This enormous diversity of receptors is generated during lymphocyte development before the cells encounter antigen. The majority of these specificities will never be needed, but it is only by this means that the immune system maintains its full potential to respond. This book examines how lymphocytes generate their antigen receptors starting from a limited number of germline genes and how the receptors on T cells and B cells actually bind and recognize antigen.

The MHC is also directly involved in the processes of T cell immune recognition and MHC molecules are also extremely polymorphic. This again is presumably related to the need to be able to present many different antigens effectively. Unlike the antigen receptors, however, the polymorphism of MHC molecules is maintained in the germline MHC genes. Different MHC haplotypes vary in their ability to present different antigens and for this reason MHC genes control immune responsiveness, acting at the level of T cell antigen recognition. The processes discussed here have far reaching effects on the operation of the immune system. For example, the ability to generate receptors which can recognize any antigen means that there must be specific mechanisms to prevent auto-immune reactions. Furthermore, the development of separate antigen receptors by T and B cells underlies the dual recognition and the cellular cooperation which occur in the development of an antibody response to T-dependent antigens. Since most antigens must be recognized both by T and B cells to produce an effective antibody response, this gives the system a high level of specificity in its ability to recognize epitopes on external antigens and distinguish them from identical epitopes on self molecules.

Ultimately the genes involved in immune recognition control the ability to mount immune responses and therefore they also determine an individual's resistance to pathogens and their susceptibility to auto-immune diseases and hypersensitivity.

2. Differences between B and T cell recognition

Although lymphocytes are functionally heterogeneous, a characteristic common to the majority of both regulatory and effector T and B cells is their capacity to recognize specific antigen. This unique property of antigen recognition is determined by the presence of specific receptors that are clonally distributed on the surface of lymphocytes, such that an individual cell recognizes only a given epitope. It is this interaction with antigen that subsequently induces expansion of the clone of lymphocytes expressing the appropriate receptor and initiates an immune response. This concept, termed the clonal selection theory (1), was originally proposed for B cells and, to hold true, necessitates the demonstration of membrane receptors capable of binding specific antigen. A variety of approaches to demonstrate antigen-binding by B cells have been reported.

Employing the direct technique of autoradiography, where the cells are incubated with radiolabelled antigen (2), the frequency of antigen-binding cells in the spleen of a primed mouse is 100- to 1000-fold that of a naive animal. However, the nature of antigen-binding receptors and their presence on the membrane of B cells is suggested by the observation that anti-Ig antisera not only bind to the surface of B cells but also inhibit the binding of antigen. A number of experiments along these lines led us to the conclusion that membrane-bound Ig is the specific receptor for antigen recognition employed by B lymphocytes (3).

Qualitatively, T cell recognition is very different in that it appears that T cells are unable to bind free antigen in the same way as B cells and need to see antigen in association with self-determinants encoded by the MHC expressed on the

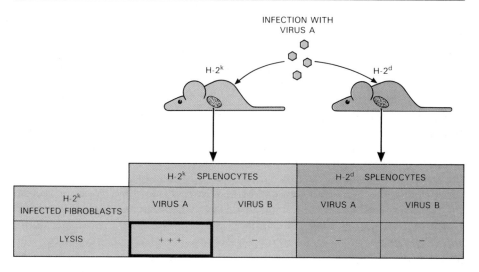

Figure 1.1. Haplotype-restricted lysis of virally infected targets. Splenocytes from H-2^k but not H-2^d mice infected with virus A are able to lyse H-2^k fibroblasts infected with virus A, but not H-2^d-infected fibroblasts. Splenocytes from neither haplotype can lyse targets infected with unrelated virus B.

Table 1.1. Comparison of T and B cell antigen recognition

B cells	T cells
1. Ligand – receptor interaction antigen – mIg	1. Ternary complex formation antigen – MHC – TCR
2. Direct binding of free antigen	2. TCR unable to bind free antigen
3. Respond to any protein	3. Response to protein restricted by MHC binding
4. Antigen epitope recognized (i) Conformational—non linear (ii) Sterically exposed (iii) ~ 15 amino acids contact	4. Antigen epitope recognized (i) Denatured-peptide fragment linear (ii) Sterically hindered (iii) 8 – 12 amino acids

surface of accessory cells, forming a trimolecular complex. Two elegant sets of experiments illustrate this requirement of T cells in the recognition of antigen. Zinkernagel and Doherty (4) noted that when spleen cells from mice of different H-2 types, infected intracerebrally with lymphocytic choriomeningitis virus, were assayed for cytolytic activity *in vitro* on a virus-infected fibroblast cell line of the H-2^k origin, only spleen cells from mice of the H-2^k haplotype specifically lysed the targets (*Figure 1.1*). Thus the co-recognition of viral antigen and MHC-encoded determinants by the T cells was required for the lysis of the targets. Similarly, in analysing the proliferative response of guinea pig T cells to soluble antigen, Shevach and Rosenthal (5) observed that activation only occurred when histocompatible macrophages and T cells were co-cultured.

The determinants (epitopes) within protein antigens recognized by B and T cells also differ (*Table 1.1*). In general, antibody raised by conventional immunization with globular protein is directed to determinants found in the native, conformationally intact molecule (6). Denatured or fragmented protein can activate T cells, thus native conformation is not a prerequisite for T cell recognition. Indeed, the majority of T cell epitopes appear to be linear. Epitopes recognized by antibody tend to correlate with highly exposed areas on the molecule. This contrasts with the T cell sites which appear to be poorly exposed (7). This may result from the steric availability of sites in that those that are exposed are preferentially degraded by proteolysis during antigen-processing by accessory cells. This would favour T cell recognition of non-conformational determinants. We have indicated that B cell recognition of antigen is that of a direct receptor/ligand interaction, whereas T cells must form a ternary complex with two ligands, namely antigen and MHC. However, to cloud the issue, recent studies on a subpopulation of human peripheral T cells suggest that the $\gamma\delta$ chain TCR complex interaction with antigen may not be MHC-restricted (8,9). Similarly, antibodies that appear to be MHC-restricted have been reported (10). As yet neither the physiological importance of these T cells nor their natural ligand is known. Nevertheless it is perhaps to be expected that T cell sub-populations will be identified that use a modification of MHC-dependent antigen recognition.

3. Structure of immunoglobulins

The basic Ig structure is that of four polypeptide chains, two identical heavy (H) and two identical light (L) chains, linked by disulphide bonds (*Figure 1.2*). Each molecule has two antigen binding sites, one on each of the Fab fragments each of which is composed of an entire L and approximately half the H chain. This model, proposed by Porter (11), was derived from the following information. Firstly, enzymatic treatment with papain cleaves the molecule into three components: two Fab fragments (which are able to bind but do not precipitate antigen, since they are univalent) and a third component, the Fc fragment, which, by crystallography and sequence, appears to be similar between antibodies, suggesting that antibody variability and heterogeneity is a function of the Fab fragments. Treatment with pepsin cleaves the Fc fragment into several small pieces, leaving the two Fab' fragments coupled by a covalent bond. This fragment, termed F(ab')$_2$, is able to both bind and precipitate antigen. Secondly, Edelman and Poulik (12) demonstrated that reduction of the disulphide bonds of intact Ig molecules revealed the presence of four chains, two of an approximate molecular weight of 53 kd and two of 22 kd (*Figure 1.2*).

Amino acid sequence analysis revealed that L chains consist of variable (VL) and constant (CL) domains, each of the order of 100–110 residues. Heavy chains consist of one VH and a variable number of CH domains, depending on the class of Ig. In the case of IgG, for example, there are three CH domains. The V and C domains themselves have little sequence homology, whereas the VH and VL, CH1 and CL, and CH2 and CH3 domains are homologous to one another. Each

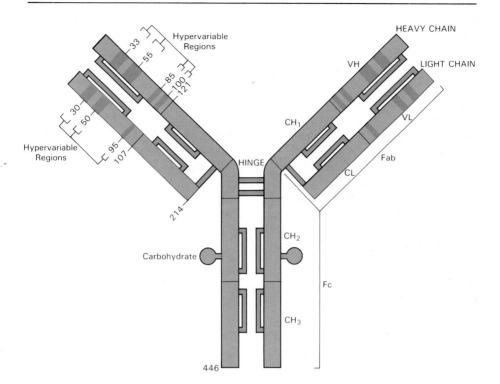

Figure 1.2.   The basic structure of an IgG molecule. The numbers indicate the amino acid residues. Orange bars show the positions of intrachain and interchain disulphide bonds. There are four domains on the heavy chains (VH, CH, etc.) and two on the light chains. Papain cleaves the molecule at the point indicated into two Fab and one Fc units.

comprises 90 – 100 amino acids with a centrally placed disulphide-linked loop (40 – 70 residues). X-Ray crystallography has revealed that domains are oval or cylindrical in shape with dimensions of 20 × 40 Å. The polypeptide chain of each domain is folded (Ig fold) such that it forms two β-pleated sheets of strands in antiparallel direction linked by a disulphide bridge (13). There are different classes and subclasses of Ig molecules which vary in structure; for example, IgG molecules have 12 domains, whereas IgM, IgD and IgE have an extra pair of constant region domains (CH4). Similarly, the number of intrachain disulphide bridges varies for individual chain types. However, our description of Ig structure will be that of IgG. Between the CH1 and CH2 domains is the hinge region of 10 – 15 residues in length (longer in IgG3) and, although showing little sequence homology between different chain types, is rich in cysteine and proline. The cysteines are involved in the disulphide bonding between the H chains and the prolines confer flexibility allowing the molecule to open and close. Furthermore, the Fab/Fc angle is variable for different Igs.

The C regions of different L and H chains of a given isotype of all individuals of the same species are encoded by the same gene complex, and this accounts

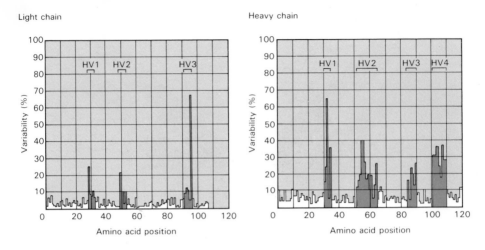

Figure 1.3. Amino acid variability at different positions in the V_L (left) and V_H (right) domains of Ig chains according to the method of Wu and Kabat (14). Hypervariable regions are indicated by shaded areas.

for the conserved amino acid sequence, although allotypic variation occurs. In contrast, the V regions of different chains are derived by recombination from numerous different variable (V), diversity (D) and joining (J) genes, resulting in the variability of the V region of individual chains. Analysis of V region amino acid sequences has shown that certain positions vary considerably more than others (Wu–Kabat plots, 14), and these are termed the hypervariable regions (*Figure 1.3*). There are at least three hypervariable regions in both V_L and V_H domains, and some consider that the part of the V domain encoded by the J gene constitutes a fourth hypervariable region. The hypervariable sequences make up approximately 25% of residues in both the V_L and V_H polypeptides. Within the hypervariable region are contact residues that bind antigen due to complementarity (complementarity-determining residues). Although by primary sequence these sites are discontinuous, they are brought together in tertiary structure to form the combining site. X-Ray diffraction studies on antigen–antibody complexes suggest that the combining site is an area between the V_L and V_H domains that varies in shape and size depending on the physico-chemical properties of the residues of the hypervariable region of the particular antibody (15,16 and Chapter 3). The more conserved residues in the V domains are termed the framework residues and constitute about 75% of the region. Indeed, some amino acids, such as tyrosine at position 86 and cysteine at positions 23 and 88, are present in all chains. Additionally, particular residues are conserved in all the different $\varkappa$ L chains (for example), but vary at equivalent positions in λ L and H chains. Finally, comparison of the framework residues of, for example, all V_λ chains reveals that they can be classified into subgroups, with different chains having the same residue at a given position (17).

Immunoglobulins are glycosylated with the carbohydrate moiety usually

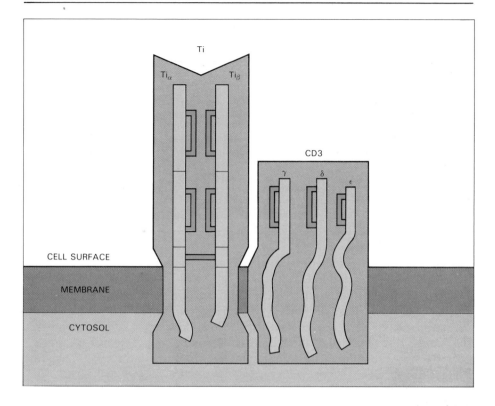

Figure 1.4. A minimal model of the TCR. Ti consists of two disulphide-bonded polypeptide chains (Ti_α and Ti_β) complexed with CD3. CD3 is composed of 5–7 chains; however, in this stylized diagram only the CD3γ, -δ and -ϵ subunits are represented.

attached to the CH2 domain of the Fc portion. The linkage is generally between the NH_2 of the asparagine and the sugar *N*-acetylglucosamine. The function of the carbohydrate component is not known, although it is thought to play a role in the secretion of some classes of Igs and in the control of catabolism.

4. Structure of the T cell receptor

Although T cells, in contrast to B cells, require co-recognition of antigen and MHC gene products, the receptors responsible for this are members of the Ig supergene family and as such share structural characteristics. The TCR was identified by antibodies with unique specificity for a particular clone of T cells (clonotypic, 18). These antibodies recognize an 80–90 kd cell surface glycoprotein containing two chains, α and β, linked by disulphide bonds (*Figure 1.4*). The α and β chains are of 43–49 and 38–44 kd molecular weight respectively, depending on the species (*Table 1.2*), and each has a polypeptide backbone of 32–34 kd and contains *N*-linked sugars. Further structural information on

Table 1.2. Components of human and murine TCR–CD3 complex

TCR(Ti)	Mol. wt (kd)	CD3	Mol. wt (kd)
1) *Human*			
Ti_α	43–49	CD3γ	25–28
Ti_β	38–44	CD3δ	20
Ti_γ	40–55	CD3ϵ	20
Ti_δ	38–42	CD3-Tp28	28
2) *Murine*			
Ti_α	40–50	CD3δ	28
Ti_β	40–50	CD3ϵ	25
Ti_γ	40–55	CD3γ	21
Ti_δ	38–42	CD3ζ	17
		CD3-p14	14

the TCR has been obtained by sequence analysis of cDNA clones encoding the α and β chains (19 and Chapter 3). The α and β chains of the TCR have two extracellular domains, one variable and the other constant, with a joining segment between. Thus, in this regard they are similar to Ig H and L chains. A transmembrane sequence links the extracellular domains to a short cytoplasmic tail. Connecting peptides of about 20 residues containing cysteines are present near the transmembrane region and may be where the interchain disulphide bridges are formed.

In addition to the conventional α/β chain heterodimer, T cells expressing a third chain, γ (55 kd), have been identified by an antibody against a peptide whose sequence was predicted from the nucleotide sequence of the γ chain gene (8). These cells have no phenotypic expression of the α/β complex. The structure of the γ chain from cDNA clone sequence analysis is similar to that of both α and β except for the absence of glycosylation sites (20). The γ chain appeared non-covalently linked to a second protein of molecular weight 40 kd, the δ chain. Interestingly, analysis of cloned CD4$^-$,CD8$^-$ thymocytes with cytolytic activity using the anti-γ chain peptide antibody identified a protein of 44 kd molecular weight complexed to a 62 kd molecule, a δ chain equivalent. The investigation of a variety of α/β chain negative cells cloned from adult and fetal blood demonstrates variable γ chain usage. Some cells expressed disulphide-linked γ chain dimers of different molecular weights, γ chains complexed with non-γ chain protein of higher molecular mass and other γ chains that may or may not be associated with a 38 kd δ chain (*Table 1.2*). This apparent heterogeneity of γ chains may result from variation in the constant region. The ability of antibodies directed against the α/β chain heterodimer to modulate antigen-specific T cell functions supports the structural evidence of this being the antigen receptor. The function of the γ and δ chains remains to be resolved. It has been proposed that γ-chain-expressing lymphocytes are T cell precursors or represent a separate lineage.

The TCR$_\alpha$, TCR$_\beta$, TCR$_\gamma$ and TCR$_\delta$ chains are expressed on the surface of lymphocytes in association with CD3 proteins (CD3γ, -δ and -ϵ) which are present on only a proportion of thymocytes but on all peripheral T cells (*Figure 1.4*; 21). In the absence of the CD3 complex, TCR chains are not expressed on the cell surface. Thus, for membrane expression of functional receptor both CD3 and TCR are required (22). It was observed that anti-CD3 antibodies under differing conditions were able to activate or inhibit antigen-reactive T cells (23). Furthermore, exposure of T cell clones to high concentrations of specific antigen or to anti-CD3 antibodies rendered the cells unresponsive to an immunogenic challenge of specific antigen (24). Taken together these experiments indicate that CD3 is functionally linked to the TCR. Immunoprecipitation studies have provided direct evidence that CD3 and TCR are complexed (25). The CD3-δ and -ε chains are both approximately 20 kd and their amino acid sequence has been derived from cDNA. The transmembrane portion of these chains contains negatively charged aspartate residues that may form salt bridges with positively charged lysines in the corresponding segment of the TCR_α and TCR_β chains. The CD3-γ chain which co-precipitates with the δ and ε chains is of higher molecular weight, 25–28 kd. The CD3 complex initially identified on human T cells is also present on murine T cells, associated with the TCR. An additional CD3 chain of 14–17 kd has been identified in the murine CD3–TCR complex. The function of CD3 appears to be signal transmission following the interaction of antigen with TCR.

5. MHC as guidance molecules

The contemporary view of MHC function is that the gene products serve as markers of self in the recognition of non-self by T cells. Firstly it is now well established that T lymphocytes are unable to bind free antigen and require to recognize antigen as a ternary complex formed with MHC gene products on the membrane of accessory cells. Furthermore, the T cells must encounter that same MHC gene product expressed by the host during their development (4). On the basis of their MHC-restriction, T lymphocytes can be separated into two main functional subpopulations, regulatory (TH/Ts) and effector cytotoxic (Tc) T cells. Helper T (TH) cells recognize antigen in association with MHC class II determinants whereas Tc tend to use MHC class I as restriction elements. While this tends to be the general rule, MHC-class-II-restricted Tc have been reported. However, MHC-restriction of the regulatory subset of suppressor T (Ts) cells is obscure, although recent information from human studies implies that the class II molecule DQw1 can function as a Ts cell-restriction element (26). Therefore, MHC gene products act as guidance molecules for T cells in the recognition of membrane-bound antigen. A variety of cell types expressing MHC class II and collectively referred to as accessory cells are able to 'process' antigen which is then presented to T cells in association with class II on the surface membrane. Initially it was thought that Tc recognition of antigen/MHC class I was different, since antibody to the specific antigen could inhibit cytolysis in certain experimental models. This could not be demonstrated for antigen-dependent activation of MHC-class-II-restricted T cells. However, evidence is now emerging from studies on viral and allo-antigen systems that Tc are able to recognize processed antigen associated with class I on the target cells in an analogous manner to TH cells. The ability of synthetic peptides in the presence of the appropriate class I or II determinants to stimulate in vitro either Tc or TH cells substantiates this (27,28). It is this complex of antigen fragment and MHC gene product that engages the antigen-specific receptor of Tc and TH cells.

The MHC class II gene products thus have a functional role in guiding TH cells in the recognition of extrinsic antigen and the subsequent initiation of an immune response. There is now increasing evidence that MHC-class-II-associated antigen recognition is required in the induction and maintenance of self-tolerance (29,30). It has been postulated that tolerance may be achieved through regulation of TH cells. The implication is that since the differentiation of immune effector function, namely antibody and Tc, requires TH, if these cells can be tolerized then autoreactivity may be circumvented. Therefore, by restricting the tissue distribution of MHC class II such that only a limited number of cell types can function as antigen-presenting cells (APC), TH need only be tolerized to self-molecules on these cells during intrathymic differentiation (31). The induction of effector function independent of TH cells would weaken this hypothesis, and recently it has been reported that Tc reactive with ectromelia virus can be generated in vivo in the absence of CD4+ TH cells (32).

6. Accessory molecules for T cells

It has become increasingly evident that cell surface glycoproteins other than the TCR – CD3 receptor complex are involved in the triggering and functional activity of T cells. Many of these proteins, termed accessory molecules (Table 1.3), appear to be primarily involved in lymphocyte adhesion (33). This is perhaps somewhat surprising since studies on, for example, lymphocyte migration have often been interpreted as suggesting that T cells are non-adhesive. Cellular contact is an essential requirement in the regulatory and effector function of T cells. Two obvious examples of this are firstly that primary activation of T cells by antigen follows after interaction with APC. Secondly, to facilitate lysis, Tc must form a conjugate with the specific target cells, although it appears

Table 1.3. Major accessory molecules for T lymphocytes

	Mol. wt (kd)	Functional effects				Target receptor
		Activation	Inhibition of Tc	Inhibition of proliferation	Inhibition of adhesion	
CD2	50	+	+	+/−	+	LFA-3
CD4	55	−	+	+/−	+	MHC class II
CD8	43,32	−	+	−	+	MHC class I
LFA-1	117,95	−	+	−	+	ICAM-1[a]
LFA-3	60	−	+	−	+	CD2

[a]ICAM-1 is probably not the only receptor for LFA-1.

that monoclonal antibodies to CD3 can induce MHC-unrestricted killing by CD3$^+$ cells (34). The specificity of T cells for antigen and MHC gene products implies that engaging the TCR receptor complex is sufficient for T cell activation. Indeed, anti-CD3 and TCR antibodies are able to inhibit Tc-mediated lysis. However, these antibodies fail to inhibit Tc/target cell conjugation (35), though antibodies reactive with LFA-1 (leukocyte functional antigen 1) and capable of abrogating both T cell proliferation and Tc-mediated lysis have been reported (36). Analysis of anti-LFA-1 antibodies on TH and TC cell clones has demonstrated their modulatory effects to be variable and overall appear to be more effective on cells with TCR – CD3 of low functional affinity. The LFA-1 family of molecules are heterodimers with a common β chain of 95 kd and three forms of α chain (α_{1-3}), and are expressed on the majority of leukocytes. In contrast, LFA-2 (E-rosette receptor; T11; CD2) is a glycoprotein of 50 kd molecular weight and is present on thymocytes and peripheral T cells. By sequence analysis of cDNA the primary structure is known (37). Although anti-CD2 antibodies can inhibit T cell function, certain combinations of these antibodies are able to induce T cell proliferation independently of macrophages (38). The ligand for CD2 now appears to be LFA-3 on the target cell and its interaction with CD2 initiates cellular conjugation prior to antigen/MHC engaging the TCR complex (39). The observation that modulation of TCR – CD3 from the cell surface inhibited the ability of anti-CD2 antibodies to activate T cells although expression of CD2 was unaltered, suggests that the role of CD2 in T cell function may be more than that of an adhesion molecule. The expression of CD2 on thymic T cells and its interrelationship with TCR – CD3 may be important in the generation of self-tolerance in the thymus (40).

The CD4 (T4) antigen is expressed primarily on a subset of mature regulatory (helper/inducer) T cells whose function and antigen recognition is restricted by MHC class II gene products, although there is heterogeneity within this population, and the CD8 (T8) antigen on this T cell subset elicits MHC-class-I-restricted cytolytic/suppressor activity. This raised the possibility that a direct interaction may occur between these accessory molecules and the respective MHC gene products. The ability of some anti-CD4 but not anti-CD8 antibodies to inhibit primary mixed lymphocyte reactions and alloreactive T cell clones with specificity for MHC class II gene products adds strength to the argument that CD4 may interact with an invariant region of MHC class II (41). This view is also supported by experiments in which murine CD4$^-$ allospecific hybridoma T cells transfected with human CD4 cDNA showed an increased response when the stimulator cells had been transfected with HLA-DR (42). However, murine fibroblasts transfected with an MHC gene construct of class II $\beta1$ domain and the class I carboxyl-terminus were able to activate an alloreactive T cell clone of the appropriate specificity. The activation was abrogated by anti-CD4 antibody, and since the MHC gene construct lacked the invariant region of class II, this suggests that the interaction between CD4 and class II is complex. In contrast, there is some suggestion that CD4 can function in the delivery of negative signals directly to CD4$^+$ T cells themselves since, in the absence of accessory cell MHC class II, crosslinking with anti-CD3 antibody failed to activate the cells

(43). In support of this is the report that anti-TCR-antibodies can co-modulate CD4 and CD3 – Ti [the idiotype$^+$ (antigen-binding) portion of the TCR] from the cell surface, implying that CD4 and the TCR complex are physically associated (44). Similarly, CD8 may initiate negative signals in T cells or (the more popular view) it may interact with an invariant region of MHC class I molecules.

Both CD4 and CD8 have been cloned and sequence analysis shows that they share significant structural homology with members of the Ig supergene family (45,46). CD4 has a variable domain with β-sheets linked by a disulphide bridge similar to Ig, and a J region linked to an extracellular domain by a region with no obvious homology. The transmembrane region has striking homology with that of the β chain of MHC class II. Although there is no sequence homology in the J region or the amino-terminus of the V domains of CD4 and CD8, and despite differences in molecular weight (CD4, 55 kd; CD8, 32 kd), they do show limited homology overall.

7. Pathways of activation

The humoral immune response to antigen develops with the proliferation of antigen-specific B lymphocytes which then mature to antibody-secreting cells. The binding of antigen to membrane Ig (mIg) is necessary, but alone is not sufficient to activate B cells, and antigen non-specific factors interacting with other receptors are required for the induction of B cell proliferation and maturation (46). Many B cell responses to antigen require the cooperation of MHC class II histocompatible TH cells (cognate recognition), which adds to the multiplicity of activating signals (47). For T-dependent antigens, binding to resting B cells may induce inositol lipid hydrolysis, mobilize intracellular calcium and cause the translocation and activation of protein kinase C (PKC; 48). In contrast, B cell mitogens such as lipopolysaccharides have no effect on inositol lipid hydrolysis but activate PKC. As a result, MHC class II expression on B cells is enhanced as they enter cell cycle (G_0 to S). These events have been determined largely by studying the effects of mitogens and anti-Ig antibodies. Following internalization and 'processing', antigen complexed with MHC class II gene products engages the TH cells (49) and stimulates the release of lymphokines, of which interleukin 4 (IL-4) and B cell growth factor II (IL-5) induce B cell proliferation. Although IL-4 alone is only able to drive B cells from G_0 to G_{1a} in the presence of suboptimal concentrations of anti-IgM antibody, the signals synergize and the cells move to S phase. The second messenger for IL-4 appears to be membrane-associated PKC. The TH – B cell interaction induces further changes in the B cells, elevating levels of cAMP, which functions as a differentiation signal. This together with T-cell-derived interferon-γ (IFN$_\gamma$) or B cell differentiation factor induces the B cells to differentiate to become Ig-secreting cells (*Figure 1.5*).

Furthermore, MHC class II determinants, in addition to their role as T cell guidance molecules, appear to be involved in membrane signalling and increasing

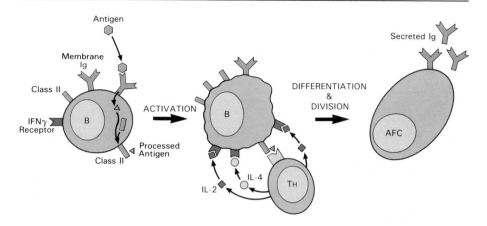

Figure 1.5. B cell activation, proliferation and differentiation. Antigen is taken up by B cell mIg, processed and presented in association with class II MHC molecules to specific TH cells. Activated B cells express receptors for IL-2 and IL-4. These receive T cell signals for differentiation and division which drive the B cells into becoming antibody-forming cells (AFCs) producing secreted Ig.

cAMP and PKC translocation to the nucleus (48). Other receptors on B cells, including the complement receptor, play a role in activation. C3, for example, following ligation appears to activate B cells and the split products themselves function as B cell growth factors (50). In contrast, occupancy of Fc receptors is a negative signal to B cells, although this can be overridden by T cell lymphokines (51).

It is now emerging that ligand – receptor interactions occurring at the membrane of T cells can, under different conditions, deliver excitatory or inhibitory signals, and the summation of these signals determines whether the outcome is to be activation. Many of the receptors intercommunicate in that down-modulation of one receptor enhances the expression of another. T cells may be activated via the TCR – CD3 complex following the binding of antigen/MHC gene product, lectin or by antibodies directed to TCR or CD3 which mimic ligand binding (52). This induces lymphokine secretion, the expression of IL-2 receptors and the cells are driven into cell cycle. An 'alternative pathway' can operate that is triggered by a combination of antibodies directed against two epitopes on CD2 (37). Exposure of T cells to anti-TCR – CD3 antibodies abrogates the mitogenic effects of CD2 stimulation; however, TCR – CD3⁻ cells can be triggered by this alternative pathway. Perturbation of TCR – CD3 and CD2 causes increased inositol lipid hydrolysis, a rise in intracellular calcium and stimulation of Na^+/H^+ exchange. The observation that calcium channel blockers inhibit TCR – CD3-induced Ca^{2+} rise has caused speculation that the CD3 complex functions as an ion transport system (53), but there is no formal proof of this and as yet the nature of the Ca^{2+} channel remains undefined since voltage clamp experiments have been unrewarding. Anti-CD3, which increases

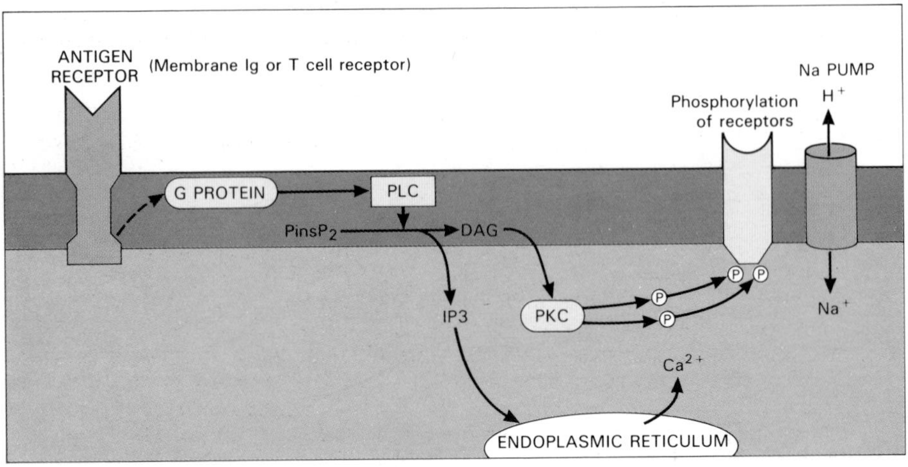

Figure 1.6. Schematic diagram of transmembrane signalling via antigen-specific receptors. G protein activates phospholipase C (PLC) which acts on phosphatidylinositol 4,5-biphosphate (PInsP$_2$) to release diacylglycerol (DAG) and inositol triphosphate (IP$_3$). These cause calcium release from the endoplasmic reticulum and activate PKC.

intracellular Ca^{2+}, and the phorbol ester phorbol myristate acetate, an activator of PKC, synergize in the production of lymphokines by enhancing transcript levels (52). PKC activation may regulate TCR–CD3 function since antigen/MHC and phorbol esters that modulate TCR–CD3 from the cell surface induce phosphorylation of CD3 (54,55). It would appear that T cell triggering requires both Ca^{2+} flux and PKC activation (*Figure 1.6*). However, there have been attempts to associate other signals with TCR–CD3-induced triggering. Recently they have concentrated on macrophage-derived IL-1. While IL-1 together with soluble anti-CD3 antibody fails to induce T cell proliferation, there are some reports of activation by insolubilized anti-CD3 and IL-1.

   Although B and T cell activation pathways use different membrane receptors and soluble mediators, they feed into common 'second messengers'.

8. Further reading

Alberts,B., Bray,D., Lewis,J., Raff,M., Roberts,K. and Watson,J.D. (1983) *Molecular Biology of the Cell*. Chapter 17, Garland, New York.

Benacerraf,B. and Unanue,E.R. (1979) *Textbook of Immunology*. Williams & Wilkins, Baltimore, MD.

Paul,W.E., Fathman,C.G. and Metzger,H. (eds) (1987) *Annu. Rev. Immunol.*, **5**.

9. References

1. Burnet,F.M. (1959) *The Clonal Selection Theory of Acquired Immunity*. Cambridge University Press, Cambridge.

2. Byrt,P. and Ada,G.L. (1969) *Immunology,* **17**, 503.
3. Paul,W.E. (1973) In Porter,R.R. (ed.), *Defence and Recognition.* Butterworths University Park Press, London, Vol. 10, p. 329.
4. Zinkernagel,R.M. and Doherty,P.C. (1974) *Nature,* **248**, 701.
5. Shevach,E.M. and Rosenthal,A.S. (1973) *J. Exp. Med.,* **138**, 1213.
6. Benjamin,D., Berzofsky,J., East,I., Gurd,F., Hannum,C., Leach,S., Margoliash,E., Michael,J., Miller,A., Prager,E., Reichlin,M., Sercarz,E., Smith-Gill,S., Todd,P. and Wilson,A. (1984) *Annu. Rev. Immunol.,* **2**, 67.
7. Thornton,J., Edwards,M., Taylor,W. and Barlow,D. (1986) *EMBO J.,* **5**, 409.
8. Brenner,M.B., MacLean,J., Dialynas,D.P., Strominger,J.L., Smith,J.A., Owen,F.L., Siedman, J.G., Ip,S., Rosen,F. and Krangel,M.S. (1986) *Nature,* **322**, 145.
9. Hersey,P. and Bolhuis,R. (1987) *Immunol. Today,* **8**, 233.
10. Wylie,D.E., Sherman,L.A. and Klinman,N.R. (1982) *J. Exp. Med.,* **155**, 403.
11. Porter,R.R. (1962) In Gelhorn,A. and Hirschberg,E. (eds), *Symposium on Basic Problems in Neoplastic Disease.* Columbia University Press, New York, p. 177.
12. Edelman,G.M. and Poulik,M.D. (1961) *J. Exp. Med.,* **113**, 867.
13. Edelman,G.M. and Gall,W.E. (1969) *Annu. Rev. Biochem.,* **38**, 415.
14. Wu,T.T. and Kabat,E.A. (1970) *J. Exp. Med.,* **132**, 211.
15. Valentine,R.C. and Green,N.M. (1967) *J. Mol. Biol.,* **27**, 615.
16. Poljak,R.J., Amzel,L.M. and Phizackerley,R.P. (1976) *Prog. Biophys. Mol. Biol.,* **31**, 67.
17. Putnam,F.W. (1974) In Brent,L. and Holborow,B. (eds), *Progress in Immunology II.* Elsevier, North Holland, p. 25.
18. Meuer,S.C., Hodgdon,J.C., Hussey,R.E., Protentis,J.P., Schlossman,S.F. and Reinherz,E.L. (1983) *J. Exp. Med.,* **158**, 988.
19. Hedrick,S.M., Cohen,D.I., Nielsen,E.A. and Davis,M. (1984) *Nature,* **308**, 149.
20. Saito,H., Kranz,D.M., Takagaki,Y., Hayday,A.C., Eisen,H.N. and Tonegawa,S. (1984) *Nature,* **309**, 31.
21. Oettgen,H.C. and Terhorst,C. (1987) *Hum. Immunol.,* **18**, 187.
22. Weiss,A. and Stobo,J.D. (1984) *J. Exp. Med.,* **160**, 1284.
23. Chang,T.W., Kung,P.C., Gingras,S.P. and Goldstein,G. (1981) *Proc. Natl. Acad. Sci. USA,* **78**, 1805.
24. Zanders,E.D., Lamb,J.R., Feldmann,M., Green,N. and Beverley,P.C.L. (1983) *Nature,* **303**, 625.
25. Reinherz,E.L., Meuer,S.C., Fitzgerald,K.A., Hussey,R.E. and Schlossman,S.F. (1982) *Cell,* **30**, 735.
26. Hirayama,K., Matsushita,S., Kikuchi,I., Iuchi,M., Ohta,N. and Sasazuki,T. (1987) *Nature,* **327**, 426.
27. Lamb,J.R., Eckels,D.D., Lake,P., Woody,J.N. and Green,N. (1982) *Nature,* **300**, 66.
28. Townsend,A.R.M., Rothbard,J., Gotch,F.M., Bahadur,G., Wraith,D. and McMichael,A.J. (1986) *Cell,* **44**, 959.
29. Groves,E.S. and Singer,A. (1983) *J. Exp. Med.,* **158**, 1483.
30. Lamb,J.R. and Feldmann,M. (1984) *Nature,* **308**, 72.
31. Cowing,C. (1985) *Immunol. Today,* **6**, 72.
32. Buller,R.M.L., Holmes,K.L., Hugin,A., Fredrickson,T.N. and Morse,H.C. (1987) *Nature,* **328**, 77.
33. Martz,E. (1987) *Hum. Immunol.,* **18**, 3.
34. Bolhuis,R.L.H. and Van de Griend,R.J. (1985) *Cell Immunol.,* **93**, 46.
35. Spits,H., Van Schooten,W., Keiser,H., Van Seventer,G., Van De Rijn,M., Terhorst,C. and De Vries,J.E. (1986) *Science,* **232**, 403.
36. Davignon,D., Martz,E., Reynolds,T., Kurzinger,K. and Springer,T.A. (1981) *J. Immunol.,* **127**, 590.
37. Sewell,W.A., Brown,M.H., Dunne,J., Owen,M.J. and Crumpton,M.J. (1986) *Proc. Natl. Acad. Sci. USA,* **83**, 8718.

38. Meuer,S.C., Hussey,R.E., Fabbi,M., Fox,D., Acuto,O., Fitzgerald,K.A., Hodgdon, J.C., Protentis,J.P., Schlossman,S.F. and Reinherz,E.L. (1984) *Cell*, **36**, 897.
39. Shaw,S., Ginther-Luce,G.E., Quinones,R., Gress,R.E., Springer,T.A. and Sanders,M.E. (1986) *Nature*, **323**, 262.
40. Reinherz,E.L. (1985) *Immunol. Today*, **6**, 75.
41. Biddison,W.E., Rao,P.E., Talle,M.A., Goldstein,G. and Shaw,S. (1983) *J. Immunol.*, **131**, 152.
42. Gay,D., Maddon,P., Sekaly,R., Talle,M.A., Godfrey,M., Long,E., Goldstein,G., Chess,L., Axel,R., Kappler,J. and Marrack,P. (1987) *Nature*, **328**, 626,
43. Bank,I. and Chess,L. (1985) *J. Exp. Med.*, **162**, 1294.
44. Saizawa,K., Rojo,J. and Janeway,C.A. (1987) *Nature*, **328**, 260.
45. Maddo,P.J., Littman,D.R., Godfrey,M., Maddon,D.E., Chess,L. and Axel,R. (1985) *Cell*, **42**, 93.
46. Littman,D.R., Thomas,Y., Maddon,P.J., Chess,L. and Axel,R. (1985) *Cell*, **40**, 237.
47. Melchers,F. and Anderson,J. (1986) *Annu. Rev. Immunol.*, **4**, 13.
48. Sprent,J. (1978) *Immunol. Rev.*, **42**, 103.
49. Cambier,J.C., Justement,L.B., Newell,M.K., Chen,Z.Z., Harris,L.K., Sandoval,V.M., Klemsz, M.J. and Ranson,J.T. (1987) *Immunol. Rev.*, **95**, 37.
50. Lanzavecchia,A. (1985) *Nature*, **314**, 537.
51. Klaus,G.G.B. and Humphrey,J.H. (1986) *Immunol. Today*, **7**, 163.
52. StC.Sinclair,N.R. and Panoskaltsis,A. (1987) *Immunol. Today*, **8**, 76.
53. Weiss,A., Imboden,J., Hardy,K., Manger,B. and Terhorst,C. (1986) *Annu. Rev. Immunol.*, **4**, 593.
54. Oettgen,H.C., Terhorst,C., Cantley,L.C. and Rosoff,P.M. (1985) *Cell*, **40**, 583.
55. Cantrell,D.A., Davis,A.A. and Crumpton,M.J. (1986) *Proc. Natl. Acad. Sci. USA*, **82**, 8158.

2

Immunoglobulin genetics and the generation of antibody diversity

1. Introduction

The means by which the immunoglobulin (Ig) system generates diversity in the B cell immune response to antigen has become clear in recent years as a result of detailed molecular genetic analyses of the Ig genes. Such studies have yielded the genetic mechanisms that generate diversity as well as the ways in which diversity may be increased through somatic and evolutionary processes. These mechanisms not only generate an enormous amount of diversity in a genetically economical way but also incorporate great flexibility into the B cell response to antigen.

A detailed knowledge of Ig molecular genetics has also explained in part several characteristics of B cell development. Thus, the molecular basis for the switch of Ig isotype from membrane-associated to secreted forms during differentiation has become clear. The large (at least 1000-fold) increase in Ig heavy (H) chain mRNA levels during development has been at least partially explained by studies on the regulatory elements controlling Ig gene expression.

Immunoglobulin genes comprise three separate loci which encode the H chain and the $\varkappa$ and λ L chains (*Table 2.1*). Each cluster contains both variable and constant region genes.

Table 2.1. Chromosomal location of Ig genes

Gene	Chromosomal location	
	Mouse	Human
IgH	12F1	14q32
Ig$_\varkappa$	6C2	2p12
Ig$_\lambda$	16	22q11

17

MOUSE IgH GENES

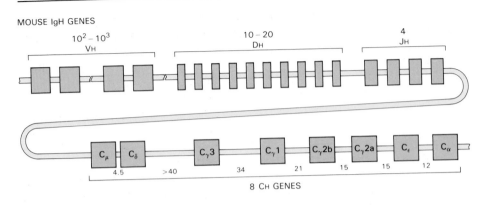

Figure 2.1. Organization of the mouse IgH locus. Gene segments are depicted as boxes. Line breaks refer to regions that have not been physically linked. Large figures indicate numbers of each kind of gene. Small figures give intron length in kb.

2. Heavy chain gene cluster

2.1 Variable region genes

The H chain variable region is encoded in the germline by three separate DNA segments: namely, the variable (V), diversity (D) and joining (J) gene segments (*Figure 2.1*). Together, the segments encode a V region of about 100–130 amino acids. The VH, DH, JH clusters are utilized by each of the H chain constant (CH) region genes (see Section 2.2).

The VH segments can encode the N-terminal 98 amino acids and are split into two exons. The 5′ exon encodes all but four amino acids of the leader sequence that mediates translocation of the H chain across the lipid bilayer of the rough endoplasmic reticulum during translation and which is cleaved during translocation. The remainder of the leader peptide and the major portion of the VH region (i.e. that not specified by the DH and JH segments) are encoded by an additional exon separated by a short intron of about 100 bp. For a variety of reasons the total number of VH segments is difficult to quantitate. Based on DNA sequence and Southern blotting analyses the VH segments have been subdivided into seven or eight subfamilies containing four to over 100 members, each sharing greater than 75% nucleotide sequence homology (1,2). Family members are clustered together with an average spacing between adjacent members of 10–20 kb (1,3,4). This estimate for the number of VH genes might, however, be conservative. In particular, Southern blotting analysis may in some cases considerably under-represent the size of a VH family because each hybridizing band may comprise multiple V genes. More accurate solution hybridization analysis of a murine VH gene family (J558) has indicated that blotting techniques have underestimated its size by at least 10-fold (5). A further complication is the existence of pseudogenes which will cross-hybridize with the probes used and which may comprise as many as 30% of VH genes (6). Although pseudo-V_H genes will not contribute directly to antibody diversity they may do so indirectly by providing a reservoir of diversity which can be utilized by unequal recombination or gene conversion (7). This is a major contributor to V_λ diversity in the chicken, where the λ locus contains a single V_λ and J_λ segment with a series of pseudo-V_λ genes adjacent to the functional V segment that contribute to diversity by a somatic gene-conversion-like process (8).

The D segments are characterized by variability both in sequence and in length (1 – 15 amino acids). The size and variability of human and murine D segments make an accurate estimation of their number difficult. Clusters of D gene segments have been identified in the genome by sequence analysis. These families range from single to ten members in the mouse and stretch over a distance of about 80 kb 5' to the J_H cluster (9,10). However, analysis of V_H amino acid sequences has suggested the possibility of extra, unidentified D segments, although this is complicated by the influence of N region diversity (see Section 6.1) at the V – D – J junction. Current (probably minimal) estimates of the numbers of D segments in the murine and human genomes are of the order of 10 – 20.

The J_H gene cluster comprises four segments in mouse and nine (including three pseudo-J_H) in man that encode the carboxy-terminal, 16 – 21 amino acids of the V_H region (11,12). The J_H segments lie about 7 kb upstream of the C_μ segment.

2.2 Constant region genes

The entire Ig C region locus of the mouse has been mapped and, together with the J_H region, spans about 200 kb of DNA in the order shown in *Figure 2.1* (11). The human C_H genes show a similar organization to their murine equivalents, although duplication of the $C_\gamma - C_\epsilon - C_\alpha$ region has clearly occurred (*Figure 2.2*) (13). Two pseudogenes are located within the human C_H region together with an additional (processed) C_ϵ pseudogene located on a separate chromosome.

The organization of the individual C_H genes is such that different exons encode the various structural domains of the corresponding protein (for review, see ref. 14). Thus, the C_μ and C_ϵ genes contain three and four exons, respectively, encoding the C region domains of the secreted protein. The C_γ genes each have an additional small exon, located between the first and second C_H domain exons, that encodes the hinge region. The α chain hinge region is encoded at the start of the second domain exon. The C_δ gene is organized somewhat differently from the other H chain genes in that it contains an extended hinge and lacks an internal C_H2 domain.

During the early stage of B cell development, Ig molecules are virtually exclusively membrane-associated (mIg) whereas plasma cells (the end-stage of B cell differentiation) synthesize secreted Ig (sIg). The genetic basis for these two forms resides in the exon organization of the C gene segments. Each Ig class is in principle able to exist in a secreted or membrane-bound form. For all classes except IgD, the unique secreted carboxy-terminal sequences are contiguous with the final C region domain. The unique sequence of mIg is encoded by a separate

HUMAN IgH GENES

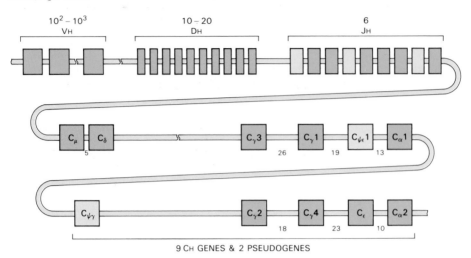

Figure 2.2. Organization of the human IgH locus. The C genes appear to have undergone gene duplication. There are two non-functional C region pseudogenes (lighter shade).

exon or exons located downstream of the last CH exon (15–17). These exons encode a hydrophobic stretch of 26 amino acids that spans the lipid bilayer of the plasma membrane and a hydrophilic cytoplasmic tail which can vary in size from three (for μm) to 28 (for γm) residues. The C_δ gene is again abnormal in that both the secreted and membrane exons are located in separate exons 3′ to the $C_\delta 3$ domain (17). Regulation of mIg versus sIg expression most probably operates at the level of RNA processing (*Figure 2.3*). Differential polyadenylation (poly A) and RNA splicing will delete the sIg carboxy-terminus and 3′-untranslated region sequences and join the membrane (M) exon(s) to the last C region domain to generate mIg mRNA (15,16).

The various M exon sequences coding for the hydrophobic transbilayer peptide show a high degree of homology. This is unusual for transmembrane sequences which generally show conservation for hydrophobicity rather than for amino acid sequence and suggests that the transmembrane segment is constrained by strong selection pressure. If the residues are placed in an α-helical conformation, one side of the helical coil is found to be especially well-conserved suggesting that intrabilayer interactions with other proteins may be involved in transmembrane signalling (18).

3. Light chain genes

3.1 λ gene locus

The λ L chain gene comprises three types of gene segment, namely V, J and C. Light chain V region genes do not contain D segments (19,20). The V segment

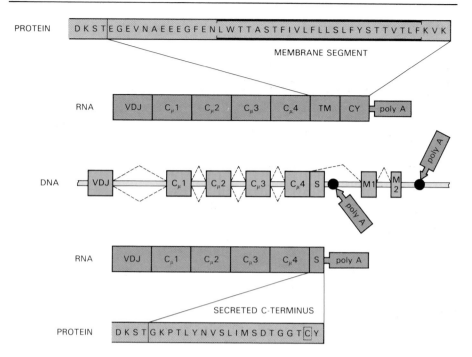

Figure 2.3.   Generation of membrane-bound (upper) and secreted (lower) forms of Ig. A rearranged μ gene is shown (DNA) with its C region exons, additional membrane exons (M1, M2) and polyadenylation sites (poly A). Alternate polyadenylation produces two types of RNA transcript which when translated produce μ chains with unique carboxy-termini. The carboxy-terminal amino acid sequences of μm or μs are represented in single letter amino acid code. The transmembrane segment is marked with bars and the cysteine residue which mediates IgM pentamer formation is boxed.

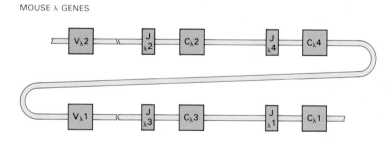

Figure 2.4.   Organization of the mouse Ig$_\lambda$ locus. Two additional C$_\lambda$ segments (not shown) have been identified within a 30 kb stretch of DNA.

in turn is split into a leader exon and one comprising the bulk of the V region coding sequence. The murine genes comprise two V gene segments, each of which is associated with two 3′ J−C segments (*Figure 2.4*). V1 joins preferentially with J1 C1 or J3 C3 and V2 with J2 C2 (21). The J4 C4 pair is defective. The

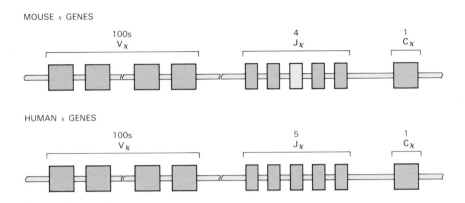

Figure 2.5. Organization of mouse and human Ig$_\varkappa$ gene loci. Note mouse has one J$_\varkappa$ pseudogene (lighter tone).

human counterpart is more complex and only partially characterized (21). The V$_\lambda$ gene pool has expanded and at least six C$_\lambda$ genes have been described. The number and positions of the human J$_\lambda$ segments are unknown. The difference in the organization of the λ locus between mouse and human may reflect the greater use of this locus in the human than in the mouse where λ comprises only 10% of the serum L chain.

3.2 *x gene locus*

The organization of the human and murine $\varkappa$ loci is similar with each comprising many V$_\varkappa$ genes and five J$_\varkappa$ genes, one of which is a pseudogene segment in the mouse, linked to a single C$_\varkappa$ segment (*Figure 2.5*). Southern blotting analysis has estimated the number of V$_\varkappa$ segments at 50 – 300 grouped into families similar to those of the VH genes, although a significant proportion (~30 – 40%) have been shown to be pseudogenes. The length of the $\varkappa$ and λ L chain V regions is relatively constant, in contrast to H chain V regions, with the V gene segment encoding amino acids 1 – 95 and the J gene segment encoding amino acids 96 – 108.

4. Rearrangement of Ig genes

4.1 *Class switching*

Switching of rearranged VHDHJH segments (see Section 4.2) between different CH genes is a characteristic of B cell development and enables the B cell compartment to provide a variety of C region-mediated effector functions associated with the same antigen-binding capacity (*Figure 2.5*). Differentiation into an Ig-secreting plasma cell is accompanied by deletion of the 5' C region genes. The mechanism of the switch is unclear; intrachromosomal deletion and unequal sister chromatid exchange may both be operational depending on the differ-

LIGHT CHAIN RECOMBINATION

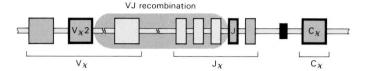

HEAVY CHAIN RECOMBINATION

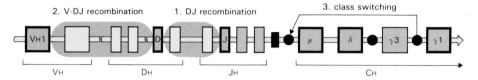

Figure 2.6. DNA rearrangements in IgL and H loci. Grey backshade indicates areas lost during recombination. Outlined genes are those which will be expressed. The closed box represents the enhancer and closed circles the switch sequences. The examples given are a $V_x2 \rightarrow J_x4$ L chain recombination, and in the H chain a $D \rightarrow J$ joining followed by $VH_1 \rightarrow DJ$. Subsequently class switching may move the $\gamma1$ (or another C gene) to the head of the gene stack to displace the initially transcribed $\mu + \delta$ segment.

entiation stage (38,39). Other mechanisms have also been implicated in more immature stages of B cell differentiation. For example, in immature B cells co-expressing IgM and IgG, differential splicing from a single primary transcript has been suggested, although it has not been ruled out that stability of μ mRNA accounts for this phenotype. Constant region switching is mediated by switch (S) regions located at the 5' side of each CH gene with the exception of C_δ (*Figure 2.5*). The S regions are tandem repetitions of a core oligonucleotide sequence and can stretch from 1 (for S_ϵ) to 10 kb (S_γ) (40,41). The way in which S sequences mediate switch recombination is unknown but may involve homologous recombination or specific proteins that recognize the various S regions (42,43).

4.2 Rearrangement of variable region genes

A functional Ig H or L chain gene is generated during B cell development by a series of rearrangement events that generate a V region coding block from the constituent gene segments (*Figure 2.6*). These rearrangement events at the three Ig gene loci occur in a well-defined order during B cell maturation (*Figure 2.7*). Studies on the order of Ig rearrangement have been greatly aided by the use of Abelson murine leukaemia virus (A-MuLV)-transformed pre-B cell lines. A-MuLV can transform immature B lymphoid cells in culture giving rise to cell lines that represent pre-B or earlier stages of B cell differentiation and some of which differentiate in culture (25,26). Analysis of these lines and of fetal liver hybridomas has provided a model for Ig gene rearrangement events during development. The initial event is H chain rearrangement which occurs at the

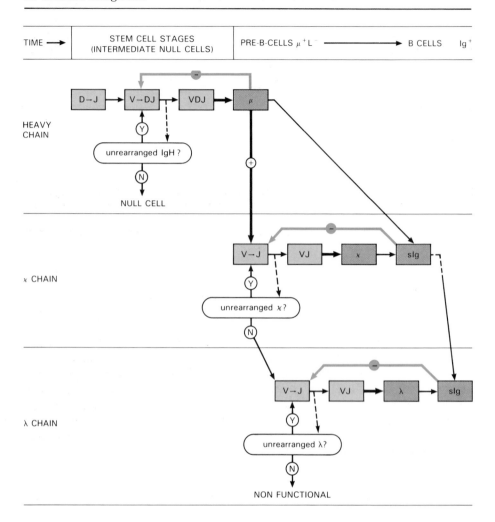

Figure 2.7. Ig gene rearrangements in B cell development. Orange boxes = events in the H, x and λ genes. Grey boxes indicate protein products. Positive and negative signals for gene rearrangement are indicated. A functional rearrangement (solid arrow) initiates the next stage. A non-functional rearrangement (dotted arrow) leaves the cell with the option of attempting another rearrangement if a gene is available (Y) or aborting the attempt to obtain a functional rearrangement of that locus if no genes are left unrearranged (N).

pre-B cell stage (27,28). This in turn encompasses two stages; the first involves rearrangement of D to JH with VH to DJH joining occurring subsequently (29). A productive (in-frame) VHDJH rearrangement on one chromosome in a pre-B cell is thought to have two consequences: (i) a 'negative' regulation event occurs in which further rearrangement at the H chain locus is suppressed (29,30) and the onset of L chain rearrangement is triggered (31). Each pre-B cell attempts to rearrange x genes before λ genes (32,33). (ii) Generation of a productively

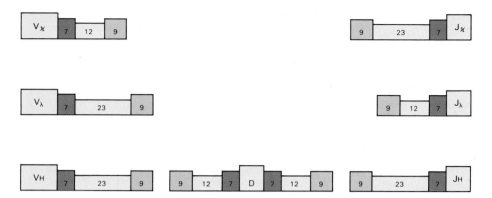

Figure 2.8. Recognition sequences involved in V gene rearrangement. The heptamer (7) and nonamer (9) sequences flanking V, D and J genes are separated by either 12 or 23 bp.

rearranged $\varkappa$ or λ gene and the consequent expression of L chain protein able to assemble into a complete Ig molecule suppresses further L chain rearrangement and results in the expression of mIg (34).

The cessation of further H or L chain rearrangement, once a productive joining has been made at the appropriate locus, effectively ensures allelic and isotypic exclusion and generates monospecific B cells. The most convincing demonstration that allelic exclusion is not simply a stocastic process resulting from a high frequency of out-of-phase joining during the rearrangement process comes from the use of mice containing functionally rearranged IgH or L transgenes. In these mice, rearrangement of endogenous genes is largely suppressed in cells expressing the transgene (35 – 37).

5. The '12 – 23' rule

Immunoglobulin V gene rearrangement is mediated by an as yet uncharacterized recombinase enzyme. The precise substrates for this enzyme and the regulation of its activity are unknown and will undoubtedly be complex. However, relatively simple sequences flanking V, D and J gene segments have been implicated as part of the recognition machinery for the recombinase (*Figure 2.8*). Short conserved sequences are found at the 3' end of all V and D gene segments and in the inverted form 5' to D and J gene segments. They consist of a conserved palindromic heptamer separated from a somewhat less conserved nonamer by a spacer whose length is either 12 ± 1 or 23 ± 1 bp (corresponding approximately to one or two turns of the double helix). The length rather than the sequence of the spacer apparently directs recombination. The empirical rule, the so-called 12 – 23 rule, that governs recombination, states that a segment with a 12 bp spacer can only recombine with a segment containing a 23 bp spacer and vice versa (44,45). The arrangement of the recognition sequences is such

RECOMBINATION WITHIN 1 CHROMATID

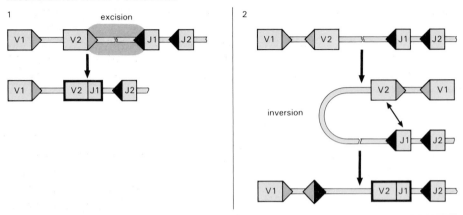

RECOMBINATION ACROSS SISTER CHROMATIDS

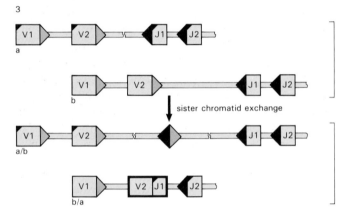

Figure 2.9. Three mechanisms for generating rearrangement. The orientation and type of the recognition sequences is shown by a black or orange triangle. Small letters a and b on (**3**) indicate the sister chromatids. In each case the recombination is between V2 and J1.

that V_H – D and D – J_H joins are permitted whereas V_H – J_H joins are forbidden (*Figure 2.8*). Clearly the 12 – 23 rule is not sufficient to predict the rearrangement process since V_H – D joins do not normally occur (see Section 4.2). Additional levels of control must therefore operate. Such controls may function by 'opening' the chromatin structure in the vicinity of the segments to be joined, allowing accessibility to the recombinase (46).

The most common mechanism by which DNA joining occurs between Ig gene segments is a deletional one; that is, the DNA between the gene segments to be joined is looped out and lost from the genome. According to this mechanism an inverted stem – loop structure stabilized by the heptamer – nonamer sequences is formed between the segments to be joined. A precise break then occurs between the V, D and J segments and the heptamer, followed by a less precise joining of the DNA segment (see Section 6.1). Direct evidence for this type of mechanism has been obtained both for Ig_κ and IgH genes (29). However, not all joining can be accounted for by such a deletional mechanism and more complex schemes such as inversion or unequal sister chromatid exchange may better explain joining in a minority of cases (*Figure 2.9*).

6. Generation of diversity

A variety of genetic mechanisms determine Ig diversity. Clearly, the multiplicity of VH and VL, DH and JH, and JL gene segments makes an important contribution. Estimates of the numbers of these gene segments in the germline have been discussed in Sections 2 and 3. In the case of the VH pool, an apparent bias in the expression of particular VH gene segments complicates estimates of the repertoire size. In pre-B cells the more JH-proximal VH genes are used preferentially (47). However, this bias is not maintained in the mature B cell population, raising the possibility that antigen selection may randomize the VH repertoire. Combinatorial association of VHDHJH and VLJL will further increase diversity, although it is possible that not every VHVL combination will allow assembly of H and L chains. Indeed, the existence of A-MuLV lines with two productive $V_\kappa J_\kappa$ rearrangements where only one of the two κ proteins will bind to the H chain argues against indiscriminate assembly of VH and VL (34,48). The potential failure to assemble is compensated for during maturation by the probable requirement of fully assembled Ig molecules for cessation of L chain gene rearrangement (see Section 4.2).

Additional mechanisms operate to extend Ig diversity considerably. These can be classified into two types, namely junctional diversity and somatic hyper-mutation.

6.1 Junctional diversity

The precise point at which V, D and J gene segments join can vary giving rise to local amino acid diversity at the junctions (49). The exact nucleotide used during joining can differ by as much as ten residues resulting in deletion of nucleotides from the ends of the V, D and J gene segments and leading to codon changes at the junctions of these segments. A consequence of this rather imprecise joining process is that two out of three rearrangements would be out-of-phase or non-productive because the joining has occurred in different translational reading frames. Analysis of the rearrangements in A-MuLV lines has borne out this predicted ratio (29).

For the VH but not the V_κ or V_λ genes additional nucleotides not encoded by either gene segment can be added at the junction between the joined gene segments during rearrangement. This process is termed N region diversity and most probably results from a template-independent addition of nucleotides by the enzyme terminal deoxynucleotidyl transferase (50).

6.2 Somatic hypermutation

This mechanism is operational at a late stage in B cell development and generates single base substitutions throughout the VH or VL gene segment and its flanking sequences. Somatic hypermutation is thought to increase the antibody affinity for an antigen during the maturation of an immune response. This can be shown most readily by studying the antibody response to haptens in which a single VH – VL combination is often utilized in the primary response. For example, in the mouse antibody response to oxazolone, mutations occur in the VH – VL pair as the immune response progresses and can be correlated with increasing antibody affinity for oxazolone (51,52). It is likely that those B cells with receptors that have higher affinities for antigen as a result of somatic hypermutation are selectively expanded, especially when antigen concentration is limiting. This process may even generate antibody-combining sites which bind antigen from germline combinations which lacked measurable affinity before somatic hypermutation, since in the secondary response to oxazolone, additional germline VH – VL combinations are detected.

6.3 VH gene replacement

An additional contribution to the generation of diversity has been observed in some CD5$^+$ (Lyl$^+$) murine pre-B cell lines and B cell lymphomas and involves the complete replacement of the functional VH gene (53,54). This replacement most probably involves an isolated heptamer sequence, homologous to that found as part of the heptamer – nonamer motif, at the 3' end of the VH coding sequence. This sequence is conserved within most VH, but not V_x or V_λ genes. Quantitatively, this mechanism probably plays a minor role in the generation of antibody diversity. Furthermore, the significance of its association with the CD5$^+$ B cells which produce the majority of auto-antibodies is unclear.

Based on the various mechanisms for generating antibody diversity, an estimate of the repertoire size can be made (*Table 2.2*). Such calculations are underestimates since they do not take into account somatic mutation, which will probably increase the repertoire by at least two orders of magnitude. Clearly, however, germline-encoded diversity alone can generate a formidable array of antibody-combining sites.

Table 2.2. The antibody repertoire

	IgH	Ig$_x$
V	1000	200
D	15	–
J	4	4
Combinatorial joining	6×10^4	8×10^2
Combinatorial association	$\sim 5 \times 10^7$	

7. Antibody – antigen interactions

Analysis of many VH and VL sequences has highlighted the role of the hypervariable regions in antigen binding. Of these, three from each chain contribute to the antibody paratope. X-Ray crystallographic analysis of Igs and latterly of antibody – antigen complexes has provided formal proof of the original suggestion that the hypervariable regions form the walls of the antigen-binding pocket. Perhaps the most elegant demonstration that the hypervariable regions provide most of the structural information necessary to confer antigen-binding specificity is the use of protein engineering to change the structure of a human myeloma Ig by transplanting the hypervariable regions of a murine anti-hapten monoclonal antibody (Mab) into the framework of the human antibody. This confers hapten-binding activity on the myeloma protein (55). Similar techniques have also been used to engineer changes in the structure of the combining site in order to increase the affinity of an antibody for its antigen (56).

7.1 Identification of epitopes

Our knowledge of how antibodies interact with antigen relies on a number of studies with structurally well-defined antigens. These include myoglobin, insulin, lysozyme and the haemagglutinin and neuraminidase of influenza A (57). In some cases it has been possible to identify epitopes composed of linear polypeptides, but more often they are formed from a number of separated amino acid residues brought together by the conformation of the antigen. The residues of a linear antigenic polypeptide which are involved in antibody binding can be readily determined by substituting different amino acids in the polypeptide and seeing whether the antibody is still able to bind to it. Within such antigenic polypeptides it is usually found that some residues are mandatory for antibody binding, some can be conservatively substituted (i.e. with a small number of structurally similar amino acids) and others may be substituted at will, without affecting the binding. Frequently, antibodies from different species or strains interact with different amino acid residues, even when they bind to the same antigenic polypeptide. This principle is also used to determine the residues involved in T cell antigen recognition which bind either to the T cell receptor or the major histocompatibility complex (MHC) molecule.

To determine the residues involved in conformational epitopes it is necessary to use intact antigens or large fragments. The earlier studies on such conformational determinants used variants of defined antigens from different species. This technique is illustrated in *Figure 2.10*, where the ability of different myoglobins to inhibit the binding of a Mab to sperm whale myoglobin (SWM) is measured (58). In this example myoglobins which vary from the sperm whale sequence at residue 83, 144 or 145 do not inhibit. This implies that the antibody paratope contacts these residues and observation of the tertiary structure of the

	Amino acid residue															Inhibition
Myoglobin	83	86	88	91	109	110	132	140	142	144	145	147	148	151	152	
Sperm whale	E	L	P	Q	E	A	N	K	I	A	K	K	E	Y	Q	+
Dwarf sperm whale	–	–	–	–	–	–	S	–	–	–	–	–	–	–	–	+
Goosebeaked whale	–	–	–	–	D	–	T	–	–	–	–	–	–	F	H	+
Killer whale	D	–	–	–	–	–	–	–	–	–	–	–	–	F	H	–
Sea lion	D	–	–	–	–	–	K	N	–	–	–	R	–	F	–	–
Human	–	I	–	–	–	C	–	–	M	S	N	–	–	F	–	–
Ox	–	V	H	E	D	–	S	N	A	E	–	–	V	F	H	–
Sheep	–	V	H	E	D	–	S	N	M	–	Q	–	V	F	–	–

Figure 2.10. Identification of myoglobin conformational epitopes. The table compares several amino acid residues of different species' myoglobin with that of sperm whale myoglobin (SWM: – = identical to SWM) and shows which myoglobins strongly inhibit binding of a Mab to SWM. Those which differ from SWM at either residue 83, 144 or 145 do not inhibit. The locations of these residues on a model of myoglobin are shown as circles (top). Residues involved in binding to another monoclonal identified similarly are shown as triangles. Based on data of Benjamin and colleagues (57).

molecule shows that the folding of the polypeptide brings these residues into proximity on the native molecule.

In vitro mutagenesis techniques now being developed will refine our knowledge of paratope and epitope contact residues. At present the structures of two Fab – antigen complexes have been solved and these have greatly extended our knowledge of B cell immune recognition.

Figure 2.11. The crystallographic structure of Fab – lysozyme, courtesy of Dr R.Poljak (59), reprinted by permission of *Science*. Copyright 1986 by the AAS. (**A**) Complex with Fab to the left, L chain uppermost and lysozyme right. (**B**) Complex separated to show complementarity and protruding residue Q 121. (**C**) Fab and lysozyme rotated forward about a 90° vertical axis to show the contact residues numbered.

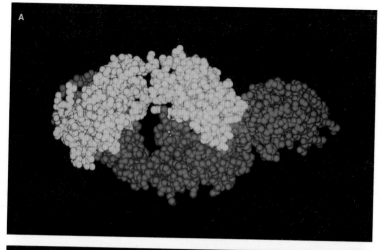

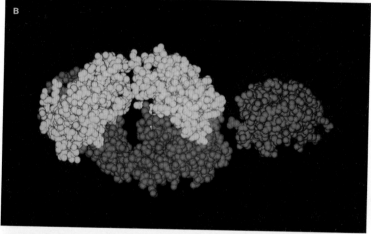

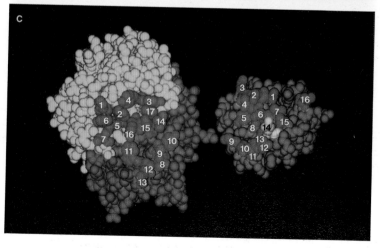

7.2 The Fab–lysozyme complex

The successful crystallization of lysozyme with Fab of the antibody D1.3 has provided insights into the nature of their association (59). The overall region of contact between antigen and antibody was about 30 × 20 Å, involving 16 lysozyme residues and 17 residues from all six antibody hypervariable loops. The majority of the contacts were made with the third hypervariable region corresponding to the D segment, an observation that may seem unsurprising when the mechanisms for generating diversity are considered. Thus both junctional and N region diversity will directly affect the structure of the third hypervariable region of H and L chains and can therefore be expected to make important contributions to binding site specificity. However, in this particular case the junctional residues of both H and L chains were not involved in antigen binding. Eleven of the 17 antibody contact residues were hydrophobic, with no electrostatic interactions evident. The contact surfaces were complementary in shape and extended beyond the paratope into surrounding regions of the Fab fragment (*Figure 2.11*). Thus the antigen combining site plays a major, although not exclusive, role in recognizing the lysozyme epitope.

7.3 The Fab–neuraminidase complex

The structure of the Fab–neuraminidase complex has revealed some additional features of antibody–antigen interaction not apparent in the lysozyme–Fab

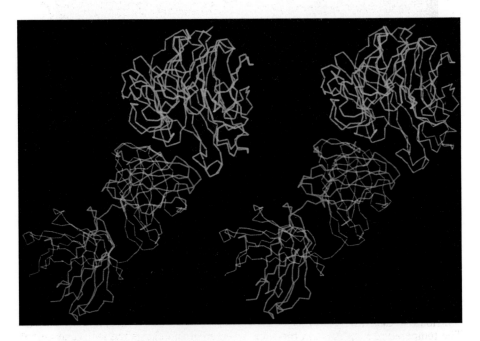

Figure 2.12. The crystallographic structure of Fab–neuraminidase, courtesy of Dr P.Colman (60). A stereo image of the C_α skeleton of Fab (blue/purple) bound to neuraminidase (green) with its active site highlighted facing forward.

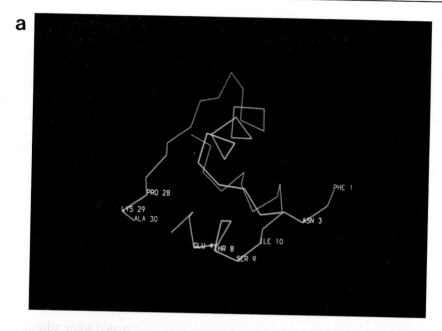

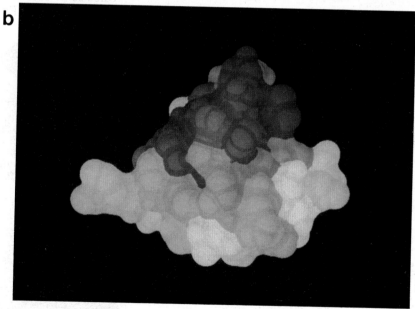

Figure 2.13. Antigenic and temperature maps of insulin courtesy of Dr J.Tainer, reproduced, with permission, from ref. 62. ©1985 by Annual Reviews Inc. Residues which contribute to four different epitopes are shown on the α-carbon skeleton of insulin (**a**). The temperature map (**b**) is in the same orientation and shows the molecular surface mobility, where the lightest areas are the most mobile. The antigenic determinants are clustered in the most mobile area of the molecule.

interaction (60). The overall area of interaction is equivalent in the two systems, with roughly the same number of interactions and all six hypervariable loops used. However, in the Fab–neuraminidase complex both the antibody and antigen undergo structural distortion, with part of the neuraminidase epitope being shifted from its position in the native structure. The position of VL is shifted relative to VH and some residues of the hypervariable regions are also moved (*Figure 2.12*). Thus, in this example there is clear evidence for an 'induced fit' model, in which conformational changes in both paratope and epitope combine to increase the energy of interaction.

8. Structure of antigens

There is much evidence to support the view that virtually the entire surface of a protein is potentially antigenic. However, the areas which are most immuno-genic depend on the species and strain of animal under investigation and it is found that most proteins have immunodominant regions. The epitopes recognized by different antibodies tend to cluster on defined parts of the antigen surface. Overlapping epitopes can be identified within these regions since antibodies to such epitopes inhibit each others binding to the antigen. It has been suggested that immunogenicity depends on a number of parameters, such as surface exposure, hydrophilicity and residue type. Recently, a role for local mobility in determining antigenicity has been proposed (61). Thus areas of protein involved in turns and loops with a high mobility are correlated with antigenicity (*Figure 2.13*; 62). A rationale for the association between antigenicity and mobility is that conformationally flexible sites can more readily adopt a complementarity to the paratope by the 'induced fit' model. Although this may be true in some cases, it is clearly not a universal rule since the region of lysozyme to which D1.3 binds is of only average mobility.

9. Further reading

Calame,K.L. (1985) *Annu. Rev. Immunol.*, **3**, 159.
Davis,D.R. and Metzger,H. (1987) *Annu. Rev. Immunol.*, **1**, 87.
Edelman,G.M. (1970) *Sci. Am.*, **223**, 34.
Leder,P. (1982) *Sci. Am.*, **246**, 102.
Tainer,J.A., Getzoff,E.D., Paterson,Y., Olson,A.J. and Lerner,R.A. (1985) *Annu. Rev. Immunol.*, **3**, 501.
Yancopoulos,G.D. and Alt,F.W. (1986) *Annu. Rev. Immunol.*, **4**, 339.

10. References

1. Brodeur,P. and Riblett,R. (1984) *Eur. J. Immunol.*, **14**, 922.
2. Dildrop,R. (1984) *Immunol. Today*, **5**, 85.
3. Kemp,D.J., Tyler,B., Bernard,O., Gough,N., Gernodakis,S., Adams,J.M. and Cory,S. (1981) *J. Mol. Appl. Genet.*, **1**, 245.

4. Bothwell,A.L.M., Paskind,M., Roth,M., Imanishi-Kari,T., Rajewsky,K. and Baltimore,D. (1981) *Cell*, **24**, 625.
5. Livant,D., Blatt,C. and Hood,L. (1986) *Cell*, **47**, 461.
6. Pech,M., Smola,H., Pohlenz,H., Straubinger,B., Gerl,R. and Zachau,H.G. (1985) *J. Mol. Biol.*, **183**, 291.
7. Baltimore,D. (1981) *Cell*, **24**, 592.
8. Reynaud,C.-A., Anquez,V., Dahan,A. and Weil,J.C. (1985) *Cell*, **40**, 283.
9. Kurosawa,Y. and Tonegawa,S. (1982) *J. Exp. Med.*, **155**, 201.
10. Wood,C. and Tonegawa,S. (1983) *Proc. Natl. Acad. Sci. USA*, **80**, 3030.
11. Shimizu,A., Takahashi,N., Yaoita,Y. and Honjo,T. (1982) *Cell*, **28**, 499.
12. Ravetch,J.V., Siebenlist,U., Korsmeyer,S., Waldmann,T. and Leder,P. (1982) *Cell*, **27**, 583.
13. Flanagan,J.G. and Rabbitts,T.H. (1982) *Nature*, **300**, 709.
14. Honjo,T. (1983) *Annu. Rev. Immunol.*, **1**, 499.
15. Alt,F., Bothwell,A., Knapp,M., Siden,E., Mather,E., Koshland,M. and Baltimore,D. (1980) *Cell*, **20**, 293.
16. Rogers,J., Early,P., Carter,C., Calame,K., Bond,M., Hood,L. and Wall,R. (1980) *Cell*, **20**, 303.
17. Word,C.J., Mushinski,J.F. and Tucker,P.W. (1983) *EMBO J.*, **2**, 887.
18. Yamawaki-Kataoka,Y., Nakai,S., Miyata,T. and Honjo,T. (1982) *Proc. Natl. Acad. Sci. USA*, **79**, 2623.
19. Tonegawa,S. (1983) *Nature*, **302**, 575.
20. Max,E.E., Seidman,J.G. and Leder,P. (1979) *Proc. Natl. Acad. Sci. USA*, **76**, 3450.
21. Eisen,H.N. and Reilly,E.B. (1985) *Annu. Rev. Immunol.*, **3**, 337.
22. Cory,S., Tyler,B. and Adams,J. (1981) *J. Mol. Appl. Genet.*, **1**, 103.
23. Bentley,D.L. and Rabbitts,T.H. (1981) *Cell*, **24**, 613.
24. Klobeck,H.G., Soloman,A. and Zachau,H.G. (1984) *Nature*, **309**, 73.
25. Rosenberg,N. and Baltimore,D. (1976) *J. Exp. Med.*, **143**, 1453.
26. Whitlock,C.A., Ziegler,S.F., Treiman,L.J., Stafford,J.J. and Witte,O.N. (1983) *Cell*, **32**, 903.
27. Levitt,D. and Cooper,M.D. (1980) *Cell*, **19**, 617.
28. Siden,E., Alt,F.W., Shinefeld,L., Sato,V. and Baltimore,D. (1981) *Proc. Natl. Acad. Sci. USA*, **78**, 1823.
29. Alt,F.W., Yancopoulos,G.D., Blackwell,T.K., Wood,C., Thomas,E., Boss,M., Coffman,R., Rosenberg,N., Tonegawa,S. and Baltimore,D. (1984) *EMBO J.*, **3**, 1209.
30. Alt,F.W. (1983) *Nature*, **312**, 502.
31. Reth,M.G., Amirati,P., Jackson,S. and Alt,F.W. (1985) *Nature*, **317**, 353.
32. Hieter,P.A., Korsmeyer,S.J., Waldman,T. and Leder,P. (1981) *Nature*, **290**, 368.
33. Coleclough,C., Perry,R., Karjalainen,K. and Weigert,M. (1981) *Nature*, **290**, 372.
34. Bernard,O., Gough,N.M. and Adams,J. (1981) *Proc. Natl. Acad. Sci. USA*, **78**, 5812.
35. Weaver,D., Costantini,F., Imanischi-Kari,T. and Baltimore,D. (1985) *Cell*, **42**, 117.
36. Rusconi,S. and Kohler,G. (1985) *Nature*, **314**, 330.
37. Ritchie,K.A., Brinster,R.L. and Storb,U. (1984) *Nature*, **312**, 517.
38. Honjo,T. and Kataoka,T. (1978) *Proc. Natl. Acad. Sci. USA*, **75**, 2140.
39. Rabbitts,T.H., Forster,A., Dunnick,W. and Bentley,D.L. (1980) *Nature*, **283**, 351.
40. Shimizu,A., Takahashi,N., Yaoita,Y. and Honjo,T. (1982) *Cell*, **28**, 499.
41. Nikaido,T., Yamawaki-Kataoka,Y. and Honjo,T. (1982) *J. Biol. Chem.*, **257**, 7322.
42. Davis,M.M., Kim,S.K. and Hood,L.E. (1980) *Science*, **209**, 1360.
43. Shimizu,A. and Honjo,T. (1984) *Cell*, **36**, 801.
44. Early,P., Huang,H., Davis,M., Calame,K. and Hood,L. (1980) *Cell*, **19**, 981.
45. Sakano,H., Maki,R., Kurosawa,Y., Roeder,W. and Tonegawa,S. (1980) *Nature*, **286**, 676.
46. Yancopoulos,G.D. and Alt,F.W. (1985) *Cell*, **40**, 271.
47. Yancopoulos,G.D., Desiderio,S.V., Paskind,M., Kearney,J.F., Baltimore,D. and Alt,F.W. (1984) *Nature*, **311**, 727.

48. Kwan,S.P., Max,E.E., Seidman,J.G., Leder,P. and Scharff,M.D. (1981) *Cell,* **26**, 57.
49. Alt,F.W. and Baltimore,D. (1982) *Proc. Natl. Acad. Sci. USA,* **79**, 4118.
50. Desiderio,S.V., Yancopoulos,G.D., Paskind,M., Thomas,E., Boss,M.A., Landau,N., Alt,F.W. and Baltimore,D. (1984) *Nature,* **311**, 752.
51. Berek,C., Griffiths,G. and Milstein,C. (1985) *Nature,* **316**, 412.
53. Even,J., Griffiths,G., Berek,C. and Milstein,C. (1985) *EMBO J.,* **4**, 3439.
53. Reth,M., Gehrmann,P., Petrac,E. and Wiese,P. (1986) *Nature,* **322**, 840.
54. Kleinfield,R., Hardy,R.R., Tarlinton,D., Dangl,J., Herzenberg,L.A. and Weigert,M. (1986) *Nature,* **322**, 843.
55. Jones,P.T., Dear,P.H., Foote,J., Neuberger,M.S. and Winter,G. (1986) *Nature,* **321**, 522.
56. Roberts,S., Cheetham,J.C. and Rees,A.R. (1987) *Nature,* **328**, 731.
57. Benjamin,D.C., Berzofsky,J.A., East,I.J., Gurd,F.R.N., Hannum,C., Leach,S.J., Margoliash,E., Michael,J.G., Miller,A., Prager,E.M., Reichlin,M., Sercarz,E., Smith-Gill,S.J., Todd,P.E. and Wilson,A.C.(1984) *Annu. Rev. Immunol.,* **2**, 67.
58. Berzofsky,J.A., Buckenmeyer,G.K., Hicks,G., Gurd,F.R.N., Feldmann,R.J. and Minna,J. (1982) *J. Biol. Chem.,* **257**, 3189.
59. Amit,A.G., Mariuzza,R.A., Phillips,S.E.V. and Poljak,R.J. (1986) *Science,* **233**, 747.
60. Colman,P.M., Laver,W.G., Varghese,J.N., Baker,A.T., Tulloch,P.A., Air,G.M. and Webster, R.G. (1987) *Nature,* **326**, 358.
61. Tainer,J.A., Getzoff,E.D., Alexander,H., Houghton,R.A., Olson,A.J., Lerner,R.A. and Hendrickson,W.A. (1984) *Nature,* **312**, 127.
62. Tainer,J.A., Getzoff,E.D., Paterson,Y., Olson,A.J. and Lerner,R.A. (1985) *Annu. Rev. Immunol.,* **3**, 339.

3

The T cell antigen receptor

1. Introduction

T and B cells possess several similarities in the way in which they recognize antigens. They both express clonally distributed receptors that must be capable of recognizing a potentially huge array of antigens. Some fundamental differences exist between the two receptor systems, the most important of which is the requirement for the T cell antigen receptor (TCR) to recognize cell-bound antigen in association with a gene product of the major histocompatibility complex (MHC) (1). The requirement for both receptors to generate extensive diversity predicts a similar overall genetic organization. The unique property of the TCR stems from the influence of the thymus in shaping the T cell repertoire (2).

Two TCRs have been defined and are referred to as $TCR_{\alpha\beta}$ and $TCR_{\gamma\delta}$ receptors. The $TCR_{\alpha\beta}$ receptor is clearly quantitatively most important in mediating MHC-restricted antigen recognition, being expressed on the majority ($>95\%$) of peripheral T cells and receptor-expressing thymocytes. Formal proof that it alone will confer antigen specificity to a T cell has been provided by DNA-mediated gene transfer of TCR_{α} and TCR_{β} genes into recipient T cells. The function of the $TCR_{\gamma\delta}$ receptor and, in particular, the identity of its ligand is, however, enigmatic.

2. TCR_{α} gene organization

The TCR_{α} gene locus (*Table 3.1*) comprises a single C_{α} gene associated with a cluster of J_{α} and V_{α} gene segments (*Figure 3.1*). As with Ig C region genes the C_{α} gene is split into exons corresponding approximately to the predicted protein domains (3,4). The external part of the α C region is encoded by two exons, one for the majority of the external region which is homologous to Ig C regions and a second for a small connecting peptide immediately adjacent to

Table 3.1. Chromosomal location of TCR genes

Gene	Chromosomal location	
	Mouse	Human
α	14 C-D	14q11
β	6 B	7q32
γ	13A2-3	7p15

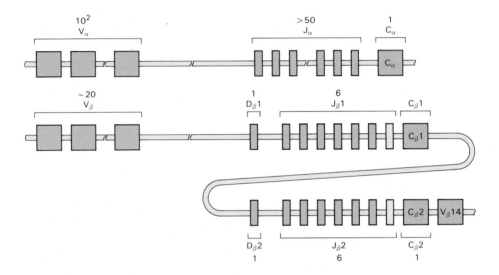

Figure 3.1. TCR$_\alpha$ and TCR$_\beta$ genomic organization. The organization of the murine genes is shown together with the numbers of the gene segments (which are approximate for V$_\alpha$, J$_\alpha$ and V$_\beta$) present in the germline. The V and C genes have a more complex exon structure than that shown here. One mouse V$_\beta$ segment is present 3' to the C$_\beta$2 gene in the opposite transcriptional orientation. No human homologue has been detected.

the plasma membrane. The transmembrane segment and short cytoplasmic tail are encoded by a third exon with the 3' untranslated region of the mRNA being located in the fourth C$_\alpha$ exon. Both C$_\alpha$ and C$_\beta$ genes lack a mechanism that generates a secreted form of the receptor.

The J$_\alpha$ locus is remarkable in two respects: namely, the large number of J$_\alpha$ gene segments and their organization over an extended genomic distance (3,4). The complement of functional J$_\alpha$ gene segments has been estimated to be at least 50, including a number of pseudogenes. The J repertoire is, therefore, considerably larger than that of the other T or B cell receptor gene families. The closest J$_\alpha$ segment is about 5 kb 5' to the C gene and the entire locus stretches over about 100 kb.

The size of the V$_\alpha$ germline repertoire is difficult to estimate with precision for the reasons discussed in Chapter 2. A combination of Southern blotting and analysis of V$_\alpha$ usage in cDNA libraries constructed from thymus or peripheral blood T cells has put the human and murine V_α repertoire at 50 – 100, divided into at least 10 different subfamilies which range in size from one to ten members (5,6).

3. TCR$_\beta$ gene organization

In mice and humans the β locus is a tandem duplication of DNA containing one D_β gene segment, several J_β gene segments and a C_β gene (*Figure 3.1*). The stretch of DNA from the beginning of the $D_\beta1$ gene to the end of the $C_\beta2$ gene spans about 20 kb.

The two C_β genes are highly homologous, with only four amino acid differences in the mouse and six in humans (7,8). These differences do not occur in corresponding residues between species and are, therefore, unlikely to mediate any functional differences between the $C_\beta1$ and $C_\beta2$ genes. Their apparently indiscriminate use by helper (TH) and cytotoxic T cell (TC) subsets is consistent with their structural similarity. Both C_β genes have an identical exon – intron organization, being split into four exons. The first exon encodes the Ig-like extracellular domain and part of a small connecting peptide that joins this domain to the plasma membrane. The remainder of the connecting peptide is encoded by a small second exon and a portion of the third exon which also encodes the transmembrane region. The fourth exon encodes a small cytoplasmic tail of 5 – 6 amino acids and the 3′ untranslated region (7,9).

There is only one known $D_\beta1$ and $D_\beta2$ segment in mouse and human, located about 600 bp 5′ to the corresponding J_β cluster (7,9). Analysis of human β chain cDNA clones has suggested that additional D_β segments may be present although these have not been detected in the germline and may equally be explained by N region diversity.

The J_β gene segments encode 15 – 17 amino acids of the V_β region. The two murine J_β clusters each contain six functional and one pseudo-J gene segments (7). The human $J_\beta1$ cluster contains six functional segments and the $J_\beta2$ cluster contains seven functional genes (10). The existence of two J_β clusters is intriguing. An attractive explanation for their evolution has been suggested by Davis (7), who has postulated that there may be a maximum number of tightly clustered J gene segments that can exist stably within the genome. If a larger number of J gene segments were required to generate a suitably sized repertoire, duplication of an entire $D_\beta – J_\beta – C_\beta$ region would be one strategy to overcome this lack of stability. An alternative strategy would be to spread out the J gene cluster over a much larger stretch of DNA, as is observed for the J_α gene segments.

The V_β gene repertoire is relatively limited, comprising about 20 members in the mouse and 50 – 100 in humans (11 – 14). A feature of the V_β gene is the small size of its subfamilies. All subfamilies in the mouse are single copy except for the $V_\beta5$ and $V_\beta8$ subfamilies, each of which contain three members. Human V_β gene segments can be divided into 15 subfamilies, the majority with one or two members. The largest human V_β subfamily comprises six members. The known murine V_β genes have been mapped to a 330 kb stretch of DNA located about 250 kb 5' to the $D_\beta1$ gene segment except for one V_β gene ($V_\beta14$), which is located about 10 kb 3' to the $C_\beta2$ gene in the opposite transcriptional orientation (15).

The strategy used by the V_β locus to encode diversity, that is, to maintain a limited repertoire of heterogeneous V_β segments, predicts that little redundancy is present within the V_β repertoire and therefore that minimal V_β polymorphism will exist. An extensive study has indeed revealed little polymorphism of the human V_β germline repertoire (16). However, several V_β polymorphisms have been detected in inbred mouse strains, the most striking of which are those resulting from the lack of about 50% of known V_β genes in several strains such as SJL (17). Surprisingly these strains fail to exhibit gross immunological abnormalities.

4. TCR$_\gamma$ gene organization

TCR$_\gamma$ gene organization is complex and in several respects reminiscent of the Ig$_\lambda$ locus (*Figure 3.2*). There are four murine C_γ genes, one of which ($C_\gamma3$) is

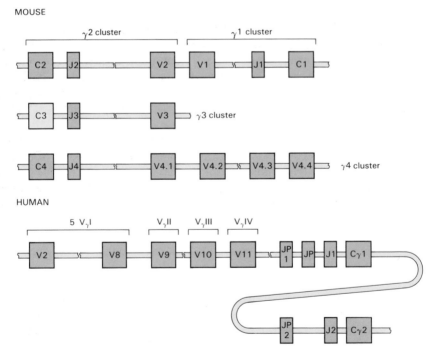

Figure 3.2. TCR$_\gamma$ genomic organization. The human and murine genes are shown. The nomenclature used for the murine γ genes is that proposed by Garman *et al.* (69). An alternative nomenclature has been used by Tonegawa (18). The exon structure of the V and C gene segments is not shown.

a pseudogene with a mutation at the splice donor site (18,19). In humans there are two known C_γ genes separated by about 16 kb (20). Each of the C_γ genes comprises three exons. The first exon encodes the extracellular domain, the second encodes most of the connecting peptide and the third encodes the remainder of the connecting peptide, the transmembrane region, the 12 amino acid cytoplasmic tail and the 3′ untranslated region. No germline D_γ segments have been identified. Although a few additional nucleotides have been found at the V – J junction they most probably originate from N region diversity. In mice each C gene is associated with a single 5′ J gene segment (18,19). The human J_γ gene segments are more extensive with three identified upstream of $C_\gamma 1$ and two upstream of $C_\gamma 2$, all of which appear to be used in T cell clones (20 – 22).

There are seven potentially functional murine V_γ genes, one associated with each of the $C_\gamma 1$, $C_\gamma 2$ and $C_\gamma 3$ loci and four linked to the $C_\gamma 4$ locus (23). Since the V_γ genes generally rearrange to adjacent J_γ gene segments, six functional $V – J_\gamma$ combinations are possible. Eight functional human V_γ genes belonging to four subgroups are located upstream of the two C_γ genes (20,21,24).

5. TCR$_x$(δ) gene organization

An investigation of apparent α gene rearrangement early in thymic ontogeny has revealed the presence of a fourth rearranging gene, called X (25). This locus comprises a C_x gene segment that exhibits structural homology to TCR and Ig C regions, together with at least one J_x and D_x gene segment (*Figure 3.3*). TCR$_x$ is located between the V_α and J_α cluster and can use at least some V_α gene segments. If every V_α has the potential to rearrange to the X locus, the potential diversity of this gene would be large.

Although formal proof is currently lacking, the TCR$_x$ gene almost certainly encodes the δ polypeptide of the receptor. Most compellingly, the TCR$_x$ gene is transcribed in cells expressing the $\gamma\delta$ receptor and is often deleted in $\alpha\beta$-expressing T cells (25). However, unequivocal identification will await the determination of δ protein sequence and comparison with the predicted sequence of the TCR$_x$ polypeptide.

MOUSE

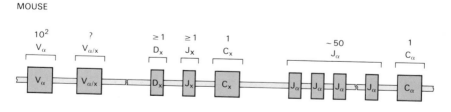

Figure 3.3. TCR$_x$ (δ) genomic organization. The relative positions of the murine C_x and C_α gene segments are shown. The segments are separated by about 80 kb of DNA. The C_x locus uses V_α-like ($V_{\alpha/x}$) genes. These genes may also rearrange to C_α.

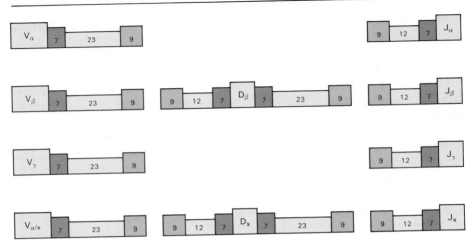

Figure 3.4. TCR recognition sequences. The heptamer (7) and nonamer (9) sequences are separated by spacer sequences of 12 and 23 bp. Adapted from ref. 26.

6. Generation of diversity

The mechanisms by which diversity is created in the T cell repertoire are, broadly speaking, similar to those utilized by B cells; that is, germline diversity coupled with gene rearrangements. The '12–23' rule first postulated to explain Ig gene joining (see Chapter 2, Section 5) also governs TCR gene rearrangement (26). All TCR V genes have two-turn recognition signals at their 3' flanking regions whereas J_α, J_β and J_γ gene segments are associated with one-turn recognition sequences (*Figure 3.4*). The $D_\beta 1$ and $D_\beta 2$ gene segments utilize a one-turn 5' recognition sequence and a two-turn 3' sequence. A consequence of this arrangement is that multiple rearrangement combinations are possible at the TCR_β locus. Thus, both $V_\beta - D_\beta - J_\beta$ and direct $V_\beta - J_\beta$ rearrangements are allowed by the '12–23' rule. Moreover, because of the arrangement of recognition sequences around the D_β genes, direct $D_\beta - D_\beta$ joining is possible. All three of these combinations have been found at the cDNA level (8,27). The $D_\beta 1$ gene segment can rearrange apparently at random both with the $J_\beta 1$ and the $J_\beta 2$ clusters (26).

As evidenced for Ig genes, TCR gene rearrangement generates considerable junctional diversity due to the imprecision of the joining process. This mechanism for the generation of diversity is particularly important for the TCR_γ genes which have a relatively restricted germline gene repertoire (20,21,23,24). As a consequence many of the rearrangements found in T cells, particularly those associated with the TCR_β and TCR_γ loci, are out of frame. Indeed, it is not unusual for a T cell line to possess three aberrant rearrangements at its TCR_β locus. A further consequence of junctional diversity in TCR_β genes is that the two D_β segments are utilized with approximately equal frequency in all three translational reading frames, in contra-distinction to Ig D gene segments, which are read in the same frame in most cases (12).

N region diversity has also been detected with V_α and probably also with V_β and V_γ genes (28,29). The enzyme terminal deoxynucleotidyl transferase which has been associated with this process is highly active in the thymus, the organ in which the TCR repertoire is generated.

Somatic hypermutation has not been detected in TCR V genes. Comparison of expressed V genes with their presumed germline counterparts has failed to detect more than isolated differences in the sequences (see e.g. ref. 29). The failure to detect somatic hypermutation does not necessarily mean, however, that this mechanism is non-existent in T cells since the majority of the functional T hybridomas and clones which have been utilized for these experiments are generated after a single immunization and either grow in an antigen-independent way or are maintained by chronic stimulation under conditions of antigen excess, conditions that are hardly optimal for the demonstration of somatic hypermutation.

7. The nature of the TCR binding site

Wu and Kabat analysis of V_α and V_β amino acid sequences has emphasized the structural similarities between TCR and Ig V regions, each containing hypervariable regions in broadly corresponding places along the sequence. Additional regions of hypervariability have been reported for V_α and V_β genes, although the relatively few V_α and V_β sequences available for these variability plots make the significance of these regions questionable (11). Further analysis of the V_α and V_β sequences has shown that although the overall homology between the various V regions is modest, many residues crucial for the Ig V region fold (including the centrally placed disulphide bond) or for the $V_H - V_L$ interaction are present in TCR V region sequences (26). The implication of these considerations is that $TCR_{\alpha\beta}$ V regions generate an analogous binding site to that of an antibody, with the hypervariable regions forming the walls of the antigen-binding pocket. The relative contributions of the three hypervariable regions may well differ between the two systems. For example, because TCR_α and TCR_β genes possess a large number of J gene segments compared with J_H, J_x and J_λ gene segments, and because the 5' ends of J gene segments are diverse, TCRs will have extra diversity within the third hypervariable region. Thus, diversity may be distributed differently within the V regions of T and B cell receptors, and this in turn may be related to the unique requirement of the TCR for MHC-linked recognition.

8. The theoretical size of the T cell repertoire

Estimates of the sizes of the T and B cell repertoires are presented in *Tables 3.2* and *2.2*. Differences in the potential T and B cell repertoires are based on differences in the extent to which the various mechanisms for generating diversity

Table 3.2. The murine $TCR_{\alpha\beta}$ repertoire

	α	β
V	100	20
D	–	2 ($\times$ 3[a])
J	50	12
Combinatorial joining	5×10^3	1.26×10^{3b}
Combinatorial association	$\sim 6 \times 10^6$	

[a]Due to the reading of $D_{\beta}1$ and $D_{\beta}2$ in all three translational frames.
[b]Assumes $D_{\beta}1$ can rearrange to $J_{\beta}1$ and $J_{\beta}2$ clusters.

are used by the two receptor systems. For example, the sizes of the VH and V_x repertoires are probably much larger than the total numbers of V_α and V_β genes; however, the J_α complement exceeds that of any of the Ig J gene repertoires. Although the $D_\beta 1$ and $D_\beta 2$ gene segments can be utilized in all three translational reading frames, their number and sequence complexity is less than for the DH gene segments. Despite the uncertainties and assumptions associated with these calculations, it is clear that the potential repertoire generated by germline, combinatorial and somatic diversity is similar for the B and T cell receptors ($\sim 10^7$ receptors).

The repertoires expressed in the periphery will of course be less than this value due to the various selective pressures exerted during B or T cell ontogeny. For both systems removal of autoreactive cells is likely to be a major influence on the repertoire. The TCR has the additional requirement for the generation of a repertoire that is restricted to self-MHC. The thymus, which is of paramount importance in T cell ontogeny, undoubtedly exerts a major influence in shaping the T cell repertoire.

9. Segregation of V_α or V_β reactivities

The identification of the TCR as a heterodimeric structure has raised the possibility that one chain may contribute predominantly MHC and the other antigen, reactivity. Analyses of V_α and V_β usage by functional T cell clones have tended to argue against this view of the TCR-binding site. For example, the same V_β gene segment was shown to be used by an I-A^d-restricted and H-2^d-restricted T cell, and two different TH cell clones, one recognizing lysozyme/I-A^b and the other cytochrome c/I-E$^{k/b}$, were found to use the same combination of V_β and J_β gene segments (26,29). Experiments such as these have failed to show a clear segregation of recognition functions with V_α or V_β. This conclusion is easily compatible with a one-binding-site view of the hetero-dimer which predicts that the relative contribution of V_α and V_β to antigen/ MHC recognition may vary with different systems in much the same way that antigen specificity in the case of immunoglobulins (Igs) may be most closely associated with the H chain, the L chain or both.

Set against this, however, are the results of several experiments that suggest a tendency for V_β to segregate with MHC recognition. An extensive study, involving sequence analysis of TCRs from a series of T cells with similar but distinct specificities for pigeon cytochrome c (p.Cytc), provided evidence for a major role for the receptor β chain in control of MHC specificity (30,31). More recently, a striking association of a particular germline V_β segment (V_β17a) with receptor specificity for $E_\beta E_\alpha$ (the MHC class II antigen I-E which comprises a two chain complex of an E_α and an E_β chain) has been demonstrated (32). Perhaps the most elegant demonstration has been the use of DNA-mediated gene transfer to confer a new and predictable MHC reactivity to a T cell hybridoma following expression of a transfected β chain gene (33). Taken together, these data suggest that a germline bias for V_β to segregate with MHC reactivity may exist and that the simple Ig-binding site view of the TCR discussed in Section 7 may be simplistic. X-Ray crystallographic analysis of the TCR and eventually of the TCR–antigen–MHC ternary complex will resolve the uncertainty.

10. Thymic development

That the thymus plays an essential role in T cell development is illustrated by the nude mouse, which lacks a thymus and has effectively no T cells, although stem cell precursors appear normal (34). Probably the most immature cells are the CD4$^-$,CD8$^-$ thymocytes, comprising about 5% of the total (35), which give rise to all the major thymic subsets (36). The majority (>90%) of thymocytes express both CD4 and CD8 molecules (37), and about 70% of these cells also express low levels of CD3 and therefore the TCR. The remaining major compartments are the CD4$^+$ and the CD8$^+$ subsets, which comprise about 5% of total thymocytes and also express high levels of CD3 (37).

In mouse the thymus is colonized at about day 11 of gestation by progenitor cells (38), which undergo a period of rapid expansion, probably driven by interleukin 2 (IL-2) (39). At about day 18 the mature CD4$^+$ and CD8$^+$ subsets appear (40). The lineages of thymocyte subpopulations have long been a matter for controversy. A reasonable sequence of development is outlined in *Figure 3.5*, although the great majority of the CD4$^+$,CD8$^+$ cells die without leaving the thymus (41). Numerous pieces of evidence support the view that CD4$^-$,CD8$^-$ cells give rise to CD4$^+$,CD8$^+$ cells, which in turn differentiate into the CD4$^+$ and CD8$^+$ subsets (42,43). This maturation scheme is undoubtedly too simple since the CD4$^-$,CD8$^-$ cells can be further subdivided according to their expression of CD5 (44), the IL-2 receptor (39), CD3 (45) and CD2 (46). It is currently debated whether these phenotypes are on a single pathway of maturation within the thymus, whether multiple pathways exist, or whether some are products of aberrant differentiation.

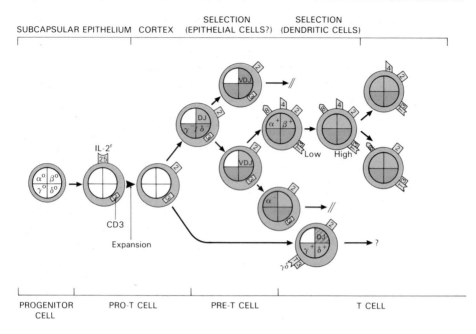

SELECTION · SELECTION
SUBCAPSULAR EPITHELIUM CORTEX (EPITHELIAL CELLS?) (DENDRITIC CELLS)

PROGENITOR
CELL
PRO-T CELL
PRE-T CELL
T CELL

Figure 3.5. A scheme for thymocyte ontogeny. A model consistent with current evidence is presented. The state of rearrangement of the four TCR loci is shown (white, unrearranged; orange, productively rearranged; grey, abortively rearranged) together with the expression of the CD3 genes and the location of the CD3 protein (intracellular or cell surface). The expression of the IL-2 receptor (IL-2^r) and the CD2, CD4 and CD8 antigens is indicated. Putative selection points for TCR$_{\alpha\beta}$-expressing T cells are given. The definitions of the cell types are: progenitor cell, blood-borne cell committed to the T cell lineage; pro-T cells, all TCR genes unrearranged; pre-T cell, one or more TCR genes rearranged, but no surface TCR; T cell, cell with surface TCR$_{\alpha\beta}$ or TCR$_{\gamma\delta}$.

11. TCR gene expression in thymocyte ontogeny

Studies of the expression of TCR genes during fetal thymocyte ontogeny have delineated a sequential series of gene expression events during gestation. The three CD3 genes, γ, δ and ϵ, are expressed early in ontogeny at around day 13; however, the CD3 protein remains cytoplasmic at this stage because the TCR genes are not expressed (47). Initial D–J rearrangements are detected at the TCR$_\beta$ locus at day 14 and complete (V–D–J) rearrangements from day 15 (48,49). Transcription of TCR$_\gamma$ and TCR$_x$ genes is also high at day 14 (48,49). TCR$_\alpha$ transcripts are, however, absent at this time and are initially detected only at day 16–17 of gestation (48,49). TCR$_\alpha$ gene expression therefore controls surface expression of the TCR receptor.

   TCR expression at the cell surface is first apparent at embryonic day 15 at which time the CD3 antigen is associated with the TCR$_{\gamma\delta}$ receptor (45). At this stage neither the CD4 nor the CD8 antigens are expressed. This receptor-positive subset represents less than 5% of the total CD4$^-$,CD8$^-$ cells and is characteristic of most, if not all, TCR$_{\gamma\delta}$-expressing T cells.

TCR$_{\alpha\beta}$ receptors are initially expressed at day 18 of gestation on CD4$^+$,CD8$^+$ cells and, at around birth (day 21), also by CD4$^+$ or CD8$^+$ thymocytes (47). A small proportion of CD4$^-$,CD8$^-$ thymocytes also express TCR$_{\alpha\beta}$ receptors at around birth, the V$_\beta$ region deriving predominantly from the V$_\beta$8 family (50,51). It has been suggested that this thymic subset represents an early stage of development of TCR$_{\alpha\beta}$-expressing T cells. However, this is difficult to reconcile with its late appearance relative to initial expression of the TCR$_{\alpha\beta}$ complex.

This sequential scheme for TCR gene expression is also a feature of adult thymocyte maturation. Double negative thymocytes and T lineage tumour lines express intracellular CD3 antigen and some express TCR$_\beta$ chains but all fail to transcribe TCR$_\alpha$ genes (46,52,53). A minor (<5%) fraction of the double negative adult thymocyte population expresses CD3 in association with the TCR$_{\gamma\delta}$ receptor (45). CD4$^+$,CD8$^+$ thymocytes and lines are heterogeneous with respect to expression of surface CD3 antigen with about 70% expressing low levels of CD3. CD3$^+$ T lineage tumours also transcribe α genes whereas CD3$^-$ lines do not (52). CD4$^+$ and CD8$^+$ thymocytes are invariably CD3$^+$ and express α and β chains.

12. Selection in the thymus

The T cell repertoire is generated in the thymus during a crucial period at or around birth (2). Adult thymectomy makes little difference to the T cell repertoire. Furthermore, since somatic hypermutation does not appear to play a significant role in the generation of diversity for TCR genes the entire repertoire must be created in the thymus during this period.

The scheme of gene rearrangement and expression events in the embryonic thymus places an early limit on the time at which the thymus shapes the T cell repertoire. The $\alpha\beta$ receptor is not expressed prior to about 18 days of gestation; selective mechanisms are, therefore, unlikely to operate prior to this time. The principle selective mechanisms that shape the repertoire initially expressed in the thymus give rise to tolerance and restriction to self-MHC (54). MHC class I and class II antigens are expressed at high levels on the stromal cells of the thymus, in particular on the thymic epithelial and macrophage/dendritic cells (55). Although cell interactions between MHC-expressing cells and TCR-expressing thymocytes must mediate selection, the crucial interactions that mediate restriction and tolerance to self-MHC have proved difficult to establish. It is likely that the macrophage/dendritic cells are the principle mediators of tolerance to self-MHC. Perhaps the most convincing evidence for this statement comes from organ culture experiments in which thymic lobes of haplotype A mice have been denuded of their haematopoietic-derived (lymphoid and macrophage/dendritic) cells by treatment with 2-deoxyguanosine and reconstituted with haplotype B progenitor cells. The lymphocyte progeny are tolerant to the donor rather than recipient haplotype (56). The mechanism by which MHC-restriction is generated is unclear, with different experimental approaches to the problem giving conflicting results (57,58). Equally compelling evidence has been presented in favour of a role for epithelial and macrophage/dendritic cells in this process. There is, however, no *a priori* reason why this function should be the sole domain of one or the other MHC-bearing cell types.

At the present time a tentative model for MHC-restriction and tolerance in the thymus would be one in which a 'window' of receptor affinities is selected. Thus, restriction would result from positive selection by interaction perhaps with epithelial cells in the thymic cortex, with the consequence that cells with affinities above a certain threshold would be expanded. Tolerance would then be mediated by a negative selection event in which macrophage/dendritic cells removed thymocytes expressing receptors with affinities above a certain threshold. Dendritic cells are well placed at the cortico-medullary junction to perform this function (55). This model, in which the selected repertoire is biased towards recognition of MHC antigens, provides a rationalization for the apparent lack of somatic hypermutation in TCR genes. With such a bias towards self-reactivity V region mutations might have a high chance of increasing the affinity above the threshold required for T cell activation and hence autoreactivity.

Clonal deletion of thymocytes expressing self-MHC-reactive receptors has recently been demonstrated directly. Cells expressing the murine $V\beta17$ gene that generates TCRs that react with a high frequency with I-E antigens are found in the thymuses of I-E-expressing mice on the $CD4^+,CD8^+$/TCR-low population but are absent from the single positive/TCR-high thymocyte subset and from peripheral T cells, demonstrating clonal elimination of these cells within the thymus (59). This observation lends strong support to the model described above.

Tolerance to non-MHC self-antigens is also mediated by the thymus around birth. The mechanism by which this form of tolerance occurs is obscure, the more so because the thymus at this stage may not be exposed to the majority of self-antigens. When this puzzle is finally solved it will undoubtedly be seen in the context of the antigen-processing requirements and protein epitopes recognized by the TCR.

13. The role of the $TCR_{\gamma\delta}$ receptor

13.1 The structure of the receptor

The $TCR_{\gamma\delta}$ has long been an enigma ever since the TCR_γ gene was first cloned and mistaken for its TCR_α counterpart. The protein product was subsequently identified by immunoprecipitation of the TCR from $CD3^+,CD4^-,CD8^-$ peripheral T cells or thymocytes using anti-CD3 antibodies under detergent conditions that stabilize the CD3 – TCR association (60). A non-$\alpha\beta$-TCR was immunoprecipitated and was shown to contain the γ protein by the use of an anti-γ antiserum raised against a synthetic peptide from the predicted protein sequence (60). An additional TCR polypeptide, termed δ, was also identified and is almost certainly the product of the fourth rearranging TCR_x gene (Section 5).

The $\gamma\delta$ receptor can exist in at least two forms in association with CD3 at the cell surface. The two chains can be associated via a disulphide link or purely by non-covalent forces depending on the γ chain constant region used (61). Further heterogeneity of the γ chain C region is generated by the use of alternate splicing to vary the numbers of repeating segments at the 5′ end of the γ C region (62). The consequences, if any, of this structural heterogeneity for $TCR_{\gamma\delta}$ function are unknown.

13.2 Lineage relationships between $\gamma\delta$ and $\alpha\beta$

There are three known T cell subsets that express the $CD3-TCR_{\gamma\delta}$ complex. The two major subsets are the $CD3^+,CD4^-,CD8^-$ cells in the thymus and periphery. A third murine population expressing this receptor is a Thy-1-positive epidermal cell of dendritic morphology that resides in the skin (63). The existence of these various subsets raises several questions concerning lineage relationships. It is not known, for example, whether peripheral $\gamma\delta$-bearing cells are precursors of their thymocyte counterparts or vice versa. It is also contentious whether a direct lineage pathway exists between $\gamma\delta$- and $\alpha\beta$-expressing thymocytes. The predominant use of different C_γ regions by $\alpha\beta$-expressing peripheral T cells (which synthesize out-of-frame $C_\gamma2$ transcripts) and $\gamma\delta$-expressing thymocytes (which express $C_\gamma1$) would appear to argue against a precursor–product relationship (64). However, definitive proof is lacking.

13.3 The function of the $\gamma\delta$-TCR

The role of the $\gamma\delta$ receptor and the interrelated questions of the nature of its ligand is a matter for intense speculation. The initial postulate by Tonegawa (48) that the γ polypeptide in combination with the TCR_β chain might recognize MHC class I antigens is incorrect since $\gamma\beta$ pairs have not been detected. However, it is likely that the $\gamma\delta$ receptor can recognize MHC antigens since a large increase in in-frame γ transcription has been observed in cDNA libraries from murine responder T cells after a mixed lymphocyte reaction (65). The use of appropriate recombinant strains should enable the responder allele(s) to be mapped.

The $\gamma\delta$ receptor can clearly be functional in the sense that it can mediate an effector response of the cell on which it is expressed. Human T cell clones expressing the $CD3-TCR_{\gamma\delta}$ receptor have been shown to exhibit natural killer (NK)-like activity (66). However, the receptor clearly cannot mediate NK activity in the majority of the large granular lymphocyte NK population since most of the cells are $CD3^-$. It is likely that the cytolytic activity described in T cells from nude mice is due to cells expressing the $\gamma\delta$ receptor since Thy-1-positive cells from the spleens of nude mice fail to transcribe the TCR_α gene but contain elevated levels ($\sim$20-fold) of in-frame γ transcripts (67).

A thymic role for the $\gamma\delta$ receptor in which interactions with its ligand result in the initiation of V_β rearrangement in a separate cell has also been postulated. This suggestion is based on the demonstration that incubation of thymus organ cultures with anti-CD3 antibodies results in the failure to rearrange V_β genes and has been interpreted on the basis of the antibody inhibiting the interactions of the $\gamma\delta$ receptor, expressed first during thymic ontogeny, with its ligand (68). If this is correct, the receptor would provide one of the many regulatory ligand – receptor interactions required for correct T cell maturation within the thymic microenvironment.

14. Further reading

Allison,J.P. and Lanier,L.L. (1987) *Immunol. Today*, **8**, 293.
Habu,S. and Okamura,K. (1984) *Immunol. Rev.*, **82**, 117.
Jerne,N.K. (1971) *Eur. J. Immunol.*, **1**, 1.
Owen,J.J.T., Jenkinson,E.J. and Kingston,R. (1983) *Annu. Immunol. (Inst. Pasteur)*, **134D**, 115.

References

1. Zinkernagel,R.M. and Doherty,P.C. (1974) *Nature*, **248**, 701.
2. Miller,J.F.A.P. and Osoba,D. (1967) *Physiol. Rev.*, **47**, 437.
3. Winto,A., Mjolsness,S. and Hood,L. (1985) *Nature*, **316**, 832.
4. Yoshikai,Y., Clark,S.P., Taylor,S., Sohn,U., Wilson,B.L., Minden,M.D. and Mak,T.W. (1985) *Nature*, **316**, 837.
5. Yoshikai,Y., Kimura,N., Toyonaga,B. and Mak,T.W. (1986) *J. Exp. Med.*, **164**, 90.
6. Arden,B., Klotz,J., Siu,G. and Hood,L. (1985) *Nature*, **316**, 783.
7. Gascoigne,N.R.J., Chien,Y.-H., Becker,D.M., Kavaler,J. and Davis,M.M. (1984) *Nature*, **310**, 387.
8. Yoshikai,Y., Anatoniou,D., Clark,S.P., Yanagi,Y., Sangster,R., van den Elsen,P., Terhorst,C. and Mak,T.W. (1984) *Nature*, **312**, 521.
9. Tunnacliffe,A., Kefford,R., Milstein,C., Forster,A. and Rabbitts,T.H. (1985) *Proc. Natl. Acad. Sci. USA*, **82**, 5068.
10. Toyonaga,B., Yoshikai,Y., Vadasz,V., Chin,B. and Mak,T.W. (1985) *Proc. Natl. Acad. Sci. USA*, **82**, 8624.
11. Patten,P., Yokota,T., Rothbard,J., Chien,Y.-H., Arai,K.-I. and Davis,M.M. (1984) *Nature*, **312**, 40.
12. Barth,R., Kim,B., Lan,N., Hunkapiller,T., Sobieck,N., Winoto,A., Gershenfeld,H., Okada,C., Hansburg,D., Weissman,I. and Hood,L. (1985) *Nature*, **316**, 517.
13. Tillinghast,J.P., Behlke,M.A. and Loh,D.Y. (1986) *Science*, **233**, 879.
14. Concannon,P., Pickering,L.A., Kung,P. and Hood,L. (1986) *Proc. Natl. Acad. Sci. USA*, **83**, 6598.
15. Malissen,M., McCoy,C., Blanc,D., Trucy,J., Devaux,C., Schmitt-Verhulst,A.-M., Fitch,F., Hood,L. and Malissen,B. (1986) *Nature*, **319**, 28.
16. Concannon,P., Gattis,R.A. and Hood,L.E. (1987) *J. Exp. Med.*, **165**, 1130.
17. Behlke,M.A., Chou,H.S., Huppi,K. and Loh,D.Y. (1986) *Proc. Natl. Acad. Sci. USA*, **83**, 767.
18. Hayday,A.C., Saito,H., Gillies,S.D., Kranz,D.M., Tanigawa,G., Eisen,H.N. and Tonegawa,S. (1985) *Cell*, **40**, 259.
19. Iwamoto,A., Rupp,F., Ohashi,P.S., Walker,L.L., Pircher,H., Joho,R., Hengartner,H. and Mak,T.W. (1986) *J. Exp. Med.*, **163**, 1203.
20. Lefranc,M.-P., Forster,A. and Rabbitts,T.H. (1986) *Nature*, **319**, 420.
21. Lefranc,M.-P., Forster,A., Baer,R., Stinson,M.A. and Rabbitts,T.H. (1986) *Cell*, **45**, 237.

22. Quertermous,T., Strauss,W.M., van Dongen,J.J.M. and Seidman,J.G. (1987) *J. Immunol.*, **138**, 2687.
23. Pelkonen,J., Traunecker,A. and Karjalainen,K. (1987) *EMBO J.*, **6**, 1941.
24. Forster,A., Huck,S., Ghanem,N., Lefranc,M.-P. and Rabbitts,T.H. (1987) *EMBO J.*, **6**, 1945.
25. Chien,Y.-H., Iwashima,M., Kaplan,K.B., Elliott,J.F. and Davis,M.M. (1987) *Nature*, **327**, 677.
26. Kronenberg,M., Siu,G., Hood,L.E. and Shastri,N. (1986) *Annu. Rev. Immunol.*, **44**, 529.
27. Rupp,F., Acha-Orbea,H., Hengartner,H., Zinkernagel,R. and Joho,R. (1985) *Nature*, **315**, 425.
28. Kavaler,J., Davis,M.M. and Chien,Y.-H. (1984) *Nature*, **310**, 421.
29. Goverman,J., Minard,K., Shastri,N., Hunkapiller,T., Hansburg,D., Sercarz,E. and Hood,L. (1985) *Cell*, **40**, 859.
30. Fink,P.J., Matis,L.A., McElligott,D.L., Bookman,M.A. and Hedrick,S.M. (1986) *Nature*, **321**, 219.
31. Sorger,S.B., Hedrick,S.M., Fink,P.J., Bookman,M.A. and Matis,L.A. (1987) *J. Exp. Med.*, **165**, 279.
32. Kappler,J.W., Wade,T., White,J., Kushmir,E., Blackman,M., Bill,J., Roehm,N. and Marrack,P. (1987) *Cell*, **49**, 263.
33. Saito,T. and Germain,R.N. (1987) *Nature*, **329**, 256.
34. Chen,W.-F., Scollay,R., Shortman,K., Skinner,M. and Marbrook,J. (1984) *Am. J. Anat.*, **170**, 339.
35. Trowbridge,I.S., Lesley,J., Trotter,J. and Hyman,R. (1985) *Nature*, **315**, 666.
36. Fowlkes,B.J., Edison,L., Mathieson,B.J. and Chused,T.M. (1985) *J. Exp. Med.*, **162**, 802.
37. Mathieson,B.J. and Fowlkes,B.J. (1984) *Immunol. Rev.*, **82**, 141.
38. Moore,M.A.S. and Owen,J.J.T. (1967) *J. Exp. Med.*, **126**, 715.
39. Jenkinson,E.J., Kingston,R. and Owen,J.J.T. (1987) *Nature*, **329**, 160.
40. Scollay,R. (1987) *Immunol. Lett.*, **15**, 171.
41. Scollay,R., Bartlett,P. and Shortman,K. (1984) *Immunol. Rev.*, **82**, 79.
42. Kingston,R., Jenkinson,E.J. and Owen,J.J.T. (1985) *Nature*, **317**, 811.
43. Smith,L. (1987) *Nature*, **326**, 798.
44. Samelson,L.E., Lindsten,T., Fowlkes,B.J., van den Elsen,P., Terhorst,C., Davis,M.M., Germain,R.N. and Schwartz,R.H. (1985) *Nature*, **315**, 765.
45. Bluestone,J.A., Pardoll,C., Sharrow,S.O. and Fowlkes,B.J. (1987) *Nature*, **326**, 82.
46. Furley,A.J., Mizutani,S., Weilbaecher,K., Dhaliwal,H.S., Ford,A.M., Chan,L.C., Molgaard, H.V., Toyonaga,B., Mak,T., van den Elsen,P., Gold,D., Terhorst,C. and Greaves,M.F. (1986) *Cell*, **46**, 75.
47. Born,W., Harris,E. and Hannum,C. (1987) *Trends Genet.*, **3**, 132.
48. Raulet,D.H., Garman,R.D., Saito,H. and Tonegawa,S. (1985) *Nature*, **314**, 103.
49. Snodgrass,H.R., Dembic,Z., Steinmetz,M. and von Boehmer,H. (1985) *Nature*, **315**, 232.
50. Fowlkes,B.J., Kruisbeek,A.M., Ton-That,M., Weston,M.A., Coligan,J.E., Schwartz, R.H. and Pardoll,J.H. (1987) *Nature*, **329**, 251.
51. Budd,R.C., Miescher,G.C., Howe,R.C., Lees,R.K., Bron,C. and MacDonald,H.R. (1987) *J. Exp. Med.*, **166**, 577.
52. Collins,M.K.L., Tanigawa,G., Kissonerghis,A.-M., Ritter,M., Price,K.M., Tonegawa,S. and Owen,M.J. (1985) *Proc. Natl. Acad. Sci. USA*, **82**, 4503.
53. Krissansen,G.W., Owen,M.J., Verbi,W. and Crumpton,M.J. (1986) *EMBO J.*, **5**, 1799.
54. Zinkernagel.R.M. (1978) *Immunol. Rev.*, **42**, 224.
55. Owen,J.J.T. and Jenkinson,E.J. (1984) *Am. J. Anat.*, **170**, 301.
56. Jenkinson,E.J., Jhittay,P., Kingston,R. and Owen,J.J.T. (1985) *Transplantation*, **39**, 331.

57. Longo,D. and Schwartz,R.M. (1980) *Nature,* **287**, 44.
58. Ron,Y., Lo,D. and Sprent,J. (1986) *J. Immunol.,* **137**, 1764.
59. Kappler,J.W., Roehm,N. and Marrack,P. (1987) *Cell,* **49**, 273.
60. Brenner,M.B., McLean,J., Dialynas,D.P., Strominger,J.L., Smith,J.A., Owen,F.L., Seidman, J.G., Ip,S., Rosen,F. and Krangel,M.S. (1986) *Nature,* **322**, 145.
61. Brenner,M.B., McLean,J., Scheft,H., Riberdy,J., Ang,S.-L., Seidman,J.G., Devlin,P. and Krangel,M.S. (1987) *Nature,* **325**, 689.
62. Littman,D.R., Newton,M., Gommie,D., Arg,S.-L., Seidman,J.G., Gettner,S.N. and Weiss,A. (1987) *Nature,* **326**, 85.
63. Koning,F., Stingl,G., Yokoyama,W.M., Yamada,H., Maloy,W.L., Tschachler,E., Shevach,E.M. and Coligan,J.E. (1987) *Science,* **236**, 834.
64. Pardoll,D.M., Fowlkes,B.J., Bluestone,J.A., Kruisbeek,A., Maloy,W.L., Coligan,J.E. and Schwartz,R.H. (1987) *Nature,* **326**, 79.
65. Jones,B., Mjolsness,S., Janeway,C.A. and Hayday,A.C. (1986) *Nature,* **323**, 635.
66. Borst,J., van de Griend,J., van Oostveen,J.W., Ang,S.-L., Melief,C.J., Seidman,J.G. and Bolhuis,R.L.H. (1987) *Nature,* **325**, 683.
67. Yoshikai,Y., Reis,M.D. and Tak,T.W. (1986) *Nature,* **324**, 482.
68. Owen,J.J.T., Owen,M.J., Williams,G.T., Kingston,R. and Jenkinson,E.J. (1988) *Immunology,* in press.
69. Garman,R.O., Doherty,P.J. and Raulet,D.H. (1986) *Cell,* **40**, 733.

4

T cell recognition of antigen and MHC gene products

1. Recognition of processed antigen

As information accumulated on the molecular structure and generation of diversity of antibody, the expectation was that T cell recognition of antigen would be similar, since they both display antigen specificity. However, B cell binding of free antigen is largely dependent on epitope conformation, whereas T cells are activated by unfolded molecules and antigen fragments in association with major histocompatibility complex (MHC) gene products.

The intracellular fate of antigen from internalization to expression on the membrane of an antigen-presenting cell (APC) remains unresolved. However, the structural properties of antigen necessary for association with MHC gene products have received much attention and their characteristics may help elucidate the mechanisms of processing. Before being recognized by T cells, globular proteins and micro-organisms, for example, must be physically altered or 'processed' by APCs such as macrophages (1). The requirement for antigen-processing was implied from the observations that antibody raised against extrinsic antigen invariably failed to inhibit T cell activation (2) and that T cells are unable to distinguish native and denatured antigen (3). Furthermore, in some instances antigen fragments were as effective in inducing T cell activation as the intact molecule (4). The treatment of APCs soon after exposure to antigen with paraformaldehyde or lysosomotropic drugs such as chloroquine will inhibit T cell activation. Therefore, binding of native antigen to the macrophage membrane is insufficient for antigen presentation and enzymatic modification (proteolysis) of the protein must occur (*Figure 4.1*; 5). This is supported by the demonstration that B cell lymphomas, after glutaraldehyde fixation, are able to present proteolytic digests of ovalbumin (OVA) but not the native or intact protein to T cell hybrids (6). Using myoglobin (MYO)-reactive T cell clones that responded equally to a 22 residue sequence either as a synthetic peptide or in the intact molecule, Berzofsky and colleagues (7) were able to investigate the antigenic requirements of T cell activation. Interestingly, an unfolded form of

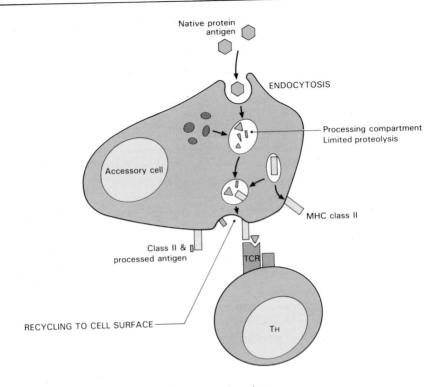

Figure 4.1. TH recognition of processed antigen.

MYO, S-methyl apoMYO, induced by chemical modification in the presence of accessory cells, was able to activate specific T cell clones as effectively as the peptide (7). Therefore, disruption of protein conformation exposing hydrophobic residues may be of equal importance to proteolysis in antigen processing (8).

That MHC-class-II-restricted helper T (TH) cells recognize processed antigen is apparent from the ability of cloned TH to respond to peptides in a variety of antigen systems. There is now compelling evidence that cytotoxic T cells (TC) also recognize processed antigen in association with MHC class I gene products. In characterizing the cytolytic response to influenza nucleoprotein (NP) using overlapping genes transfected into fibroblasts it was observed that short fragments of protein could be recognized by the TC cells (9). Interestingly, the NP signal sequence for membrane transport was unnecessary, suggesting that protein degradation may occur in the cytosol. This was tested using synthetic peptides and the results demonstrated that the TC clones recognized an epitope located within residues 365 – 379 (10). Studies on the human response to influenza virus also demonstrated the presence of MHC-class-I-restricted TC that recognize peptides (11). This applies not only to viral proteins that are expressed internally but also to proteins present as membrane components since murine TC specific for HLA-CW3 were able to lyse HLA-negative syngeneic spleen cells in the presence of the appropriate synthetic HLA peptide (12).

The region of MYO around K-140 (residues 132 – 146) is an immunodominant T cell site in the H-2^d haplotype that forms an amphipathic α-helix (7). This generates two facets, one of which is hydrophobic and the other hydrophilic, giving a polarity which allows the spatial separation of MHC and T cell receptor (TCR) interaction sites that overlap sequentially within an antigen. It is suggested that the hydrophobic face binds to the MHC gene product and the hydrophilic to the TCR (13), as amphipathic structures have natural affinity for lipid membranes. This may help to concentrate antigen and facilitate the binding of antigen to MHC molecules which is a relatively low affinity interaction. Although this hypothesis is attractive it fails to account for those T cell epitopes that occur in strands. An alternative model has been proposed by Rothbard (14) based on the linear patterns of residues in a peptide. The pattern begins with a charged residue or glycine followed by two or three consecutive hydrophobic residues and terminates with a polar amino acid. Modification of T_H or T_C epitopes by deletion of residues composing these motifs abrogated their ability to activate the relevant T cells. It may be that these patterns identify residues critical in the binding of antigen to MHC molecules or allow the peptide to adopt a conformation enabling it to engage both the MHC and TCR.

2. Direct binding of peptide and MHC molecules

Attempts to demonstrate directly the presence of antigen on the presenting cell membrane bound to MHC class II gene products have been lacking. That this complex alone is sufficient for T cell recognition is illustrated by the ability of lipid membranes containing MHC class II molecules of the appropriate specificity in the presence of antigen to generate T cell function (15). However, the observation that antigens could compete one with another at the level of APCs (16,17) provided the initial evidence that processed antigen and MHC class molecules are able to interact specifically, prior to recognition by T cells. Analysis of the supernatants of antigen-pulsed macrophages indicated the presence of antigen – MHC class II complexes (18), but formal biochemical evidence of their existence was provided by equilibrium dialysis (19). This demonstrated that affinity-purified MHC class II from a high-responder haplotype (I-A^k) in non-ionic detergent solution was able to bind specifically a peptide from hen egg lysozyme (HEL; residues 46 – 61) with a dissociation constant of about 5 μM. Binding of this peptide to non-responder class II (I-A^d) was some 10-fold less. These observations were confirmed for OVA peptide 323 – 339, which bound to I-A^d but not to I-E^d or class II of the non-responder haplotype H-2^k (20). Competition with cold OVA 323 – 339, unlike the truncated peptide 329 – 339, inhibited the binding of [^{125}I]OVA 323 – 339. This parallels the functional results in that OVA 323 – 339 is I-E^d-restricted and OVA 329 – 339 fails to activate T cells induced with OVA 323 – 339. Using OVA peptides sequentially truncated at the amino- and carboxy-termini in binding studies, the critical residues required for the binding to I-A^d could be mapped within a seven amino acid sequence 327 – 333. Substitutions at each position demonstrated that four amino acids, V-327, H-328, A-332 and E-333 (amino acid single letter code), formed contact residues with I-A^d of which only V and A were strongly implicated in binding (21). The residues H, A and E also appeared to contact the TCR. Interestingly, only non-conservative changes at any of these four positions affect binding, suggesting that the specificity of the antigen – MHC class II interaction is broad. Crosslinking the OVA – MHC class II complex with glutaraldehyde demonstrated that binding occurred between the peptide and the I-A^d α but not β chain. This is somewhat surprising since the polymorphism resides in the β chain. It is possible that antigen binds to both α and β chains but for technical reasons only crosslinks to the α chain; however, this seems unlikely since MYO (132 – 153) and haemagglutinin (111 – 120) crosslink to β and α chains respectively.

Since peptides from unrelated proteins such as influenza haemagglutinin (130 – 142) that are I-A^d-restricted also bind, they have been analysed for structural similarities with OVA 323 – 339. Indeed, a common motif could be identified that was absent in non-I-A^d-binding peptides. Furthermore, as some residues interact with both MHC class II and the TCR it would appear unlikely that the OVA 323 – 339/I-A^d complex is an amphipathic α-helix but forms a β-sheet or other extended structure.

Noting that antigen – MHC class II complexes are stable to gel-filtration allowed bound and free peptide to be separated and thus the kinetics of the association and dissociation of these complexes could be analysed (20). In solution complex formation is 5-fold slower than that of antibody – antigen interactions, at a rate of $K_a \cong 1 M^{-1} s^{-1}$, but once formed is very stable. The dissociation of peptide from MHC class II at 37°C, similar to antibody – antigen complexes, has a half life of 10 h (*Table 4.1*). The slow association rate implies that either processed antigen must be present at high concentration in the extracellular fluid or that it is localized with MHC class II in the same intracellular compartment for a relatively long period of time before exposure on the cell membrane. Alternatively, fluorescence energy transfer studies suggest that the presence of the TCR increases the amount of antigen associated with MHC class II gene products, thereby stabilizing a low affinity interaction (22). Resolving problems such as this are fundamental in understanding the mechanisms of antigen-processing.

3. Determinants recognized: antigen – TCR, antigen – MHC and MHC – TCR

As the TCR is unable to bind free antigen, indirect assays have been adopted to identify critical contact residues present in antigenic determinants. T cells of B10.A mice primed with moth cytochrome *c* (m.Cytc) peptide 88 – 103 (K at residue 99) responded weakly to the peptide analogue Q-99. Conversely, mice primed to the Q-99 analogue responded to Q-99 and not the K-99 peptide. Since the same APCs were presented to both T cell populations it suggests that the

Table 4.1. Kinetics of MHC class II – antigen interaction

	MHC class II – antigen	Hyperimmune antibody – antigen
K_a	10 M^{-1} s^{-1}	10^5 M^{-1} s^{-1}
K_d	10^{-5} – 10^{-6} M^{-1} s^{-1}	10^4 M^{-1} s^{-1}
K_D	10^{-5} – 10^{-6} M	10^{-9} – 10^{-11} M

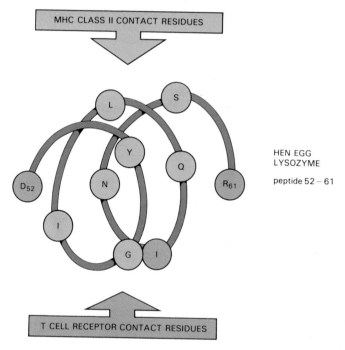

Figure 4.2. MHC class II and TCR contact residues of peptide 52 – 61 of lysozyme. MHC contact residues are in orange, TCR contact residues in brown, residues which bind to neither are grey. Residues are in single letter amino acid code. The peptide is thought to adopt a helical configuration.

effect of substitutions at residue 99 was not on antigen-processing but selected T cells of different specificities (23). Recently, Allen *et al.* (24) mapped the TCR contact residues in HEL peptide 52 – 61. Peptide analogues with A sequentially substituted at each residue were assayed for their ability to either activate T cell clones or competitively inhibit presentation of 52 – 61 (*Figure 4.2*).

Structural analysis suggested that the peptide adopted an α-helical conformation with TCR and MHC class II contact residues on opposing facets. Using a similar approach in the analysis of T cell recognition of OVA peptide 323 – 339, it appeared that three residues only interacted with TCR, while another three made contact with both TCR and MHC class II (21). In contrast to HEL, the conformation of OVA 323 – 339 was postulated to be that of a β-sheet 'sandwiched' between the TCR and MHC class II (*Figure 4.3*).

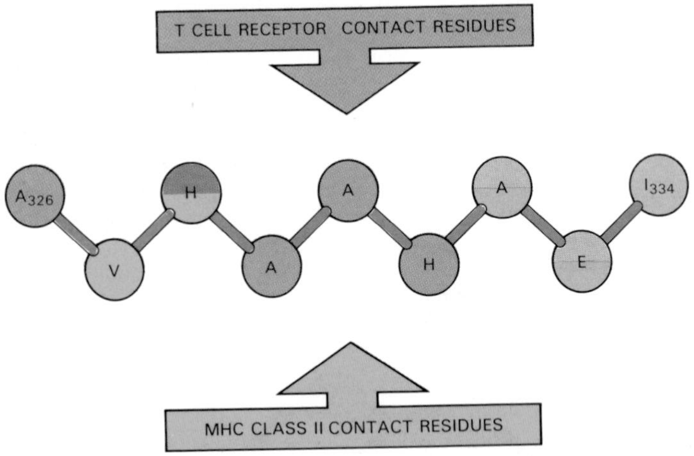

Figure 4.3. MHC class II and TCR contact residues of chicken OVA residues 326 – 334. The peptide adopts a β-sheet configuration within the trimolecular complex. Residues colour coded as in *Figure 4.2*.

The presence of a functional 'subsite' in antigenic determinants that interacts with MHC gene products is suggested by a number of functional experiments. It was discovered that certain B10.A T cell clones induced with pigeon cyto-chrome *c* (p.Cytc) while able to recognize m.Cytc in association with either B10.A or B10.A(5R) MHC class II products but no other haplotype, responded to p.Cytc only in association with B10.A (25). Thus, with the TCR constant, the altered pattern of T cell reactivity was attributable to variation in the antigen – MHC interaction. The specificity of this interaction is supported by the observation that related peptides can compete for presentation at the level of the APC (16,17). Recent experiments demonstrated that analogues of a peptide from λ cI protein and peptides from unrelated proteins but presented by the same restriction element could competitively inhibit the activation of T cell clones reactive with λ cI repressor protein (26). Analysis of the primary sequence of the λ cI peptide and other peptides that interact with the same MHC molecule (I-E^d) showed they contained similar amino acid patterns that are also present in the hypervariable region of I-E^d (27). This was interpreted to suggest that these residues in the MHC molecule bind to a complementary site in the α chain and for extrinsic antigen to bind it must displace this internal interaction within the MHC chains. From both direct binding and competitive inhibition assays it appears that only a few antigen residues interact with a single binding site of broad specificity on MHC class II molecules (*Figure 4.4*).

From studies on allospecific T cell recognition it is apparent that the TCR must interact with MHC molecules, but whether or not the mechanisms of this interaction differ from those controlling the recognition of extrinsic antigen is controversial. Nevertheless, it appears that the formation of a stable trimolecular complex between TCR, antigen and MHC class II is influenced by the interaction

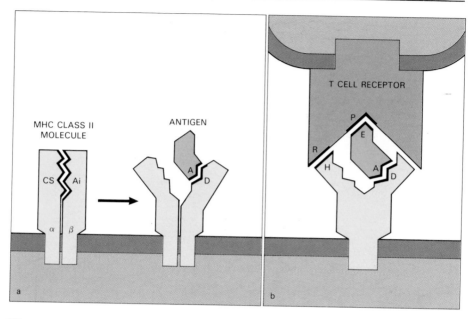

Figure 4.4. Models of the interaction between antigen – MHC – TCR. A = the agretope. Ai = internal agretope that in the absence of antigen binds the complementary site (CS) on Ia α chain. D = desetope, the site on Ia which contacts antigen. E = epitope, the site on the antigen which contacts the TCR. H = histotope, the site on MHC which contacts the TCR. P = paratope, the site on the TCR which contacts antigen and R = restitope, the site on the TCR which contacts MHC.

between TCR and MHC class II. Functional evidence for this comes from the analysis of the B10.A and B10.A(5R) T cell responses to p.Cytc and m.Cytc (28). The B10.A responds to both m.Cytc and p.Cytc but the B10.A(5R) only to moth; however, B10.A(5R) APCs can present p.Cytc to B10.A T cells. Furthermore, approximately 50% of these B10.A(5R) T cells respond to m.Cytc in association with B10.A-presenting cells. This suggests that T cells from the two strains with similar epitope specificity appear capable of recognizing a common site on the two different MHC class II molecules (MHC – TCR interaction). The degeneracy of the T cell response observed on B10.A (E_β^k) and B10.A(5R) (E_β^b) but not B10S(9R) (E_β^s) APCs may result from the presence of identical residues in the hypervariable sequence (68 – 75) of the β1 domain of E_β^k and E_β^b. Additional evidence suggesting that residues located in the third hypervariable β1 domain may interact directly with the TCR has come from the analysis of amino acids at positions 67, 70 and 71 in A_β^b of the bm12 mutation. Substitution of any one of these residues in the parent b haplotype with that from the corresponding position of the bm12 abrogated the response of the majority of T cells irrespective of their antigen specificity (29). The implication of this is that the effect occurs at the level of the MHC class II interaction with the TCR rather than antigen.

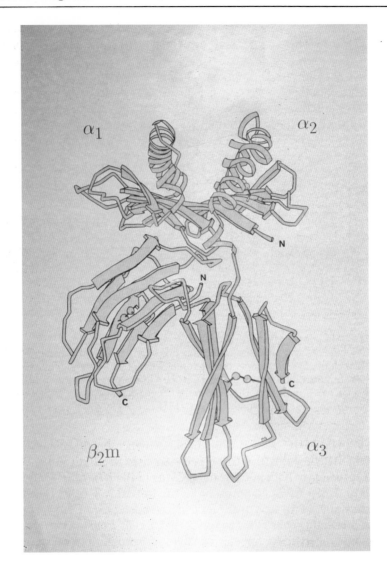

Figure 4.5. Polypeptide chain structure of HLA-A2, by courtesy of Dr D.C.Wiley, reprinted by permission from *Nature* (33). Copyright © Macmillan Magazines Limited.

Residues 63–78 are probably on the external surface of the $\beta1$ domain and together with the α chains of MHC class II may form direct contacts with the TCR (30). Taken together these results imply that MHC class II molecules contain distinct functional sites for antigen and TCR binding. Recent experiments demonstrate that the large majority of Vβ17a$^+$ T cell hybridomas react with I-E of all haplotypes and since these class II molecules have highly conserved α chains but polymorphic β chains it suggests that Vβ17a$^+$ receptors bind to the I-E$_\alpha$ chain (31). The additional hypervariable regions of V$_\beta$ of the TCR may be located outside the antigen binding site if comparable with immunoglobulin, although insufficient sequences have been analysed to confirm this. However, if these additional sequences do exist it is tempting to speculate that they form the restitope; the site on the TCR which contacts MHC.

The experiments that we have discussed above provide some insights as to the physical interactions that occur between MHC and TCR molecules in the recognition of antigen. Now that the crystal structure of the MHC class I, HLA-A2 (33,34) has been determined it will be possible to resolve the physical basis of these interactions. From the three-dimensional structure it is evident that there is a single antigen recognition site present as a groove located on the top surface of the molecule (*Figure 4.5*). The sides of this groove are formed by two long α-helices of the $\alpha1$ and $\alpha2$ domains, and the floor by β-strands also originating from the $\alpha1$ and $\alpha2$ domains. The polymorphic amino acids of the different class I alleles are located almost entirely in the antigen binding site. The residues that project upward and away from the α-helices are potentially those that form contacts with the TCR, whereas those that point inwards from the helices together with the residues of $\alpha1 - \alpha2$ β-sheet on the floor of the groove affect the binding of antigen peptides (34). Since the overall size of the antigen recognition site is approximately 25 Å long, 10 Å wide and 11 Å deep it could accommodate peptide antigens of 8 – 20 amino acids if their conformation is that of an α-helix or extended chain (21,24). Interestingly, from the electron density map it appeared that during crystallization of HLA-A2, peptide(s) had bound in the proposed antigen recognition site (34). Although their origin is not known, it is tempting to speculate that peptides, possibly from 'self', always occupy the binding sites of MHC molecules and that they have to be displaced by foreign antigen before it can induce an immune response (27). As regards the structure of MHC class II molecules, modelling the $\alpha1$ and $\beta1$ domains to the same structure as HLA-A2 also positions the polymorphic residues in the proposed antigen recognition site; however, conformation of this must await the crystalline structure.

4. Models for allorecognition

When lymphocytes from histo-incompatible donors of the same species are cultured together the T cells from each individual proliferate in response to the foreign MHC gene products. These allogeneic responses are directed to both MHC class I and class II molecules. Whether the mechanism of T cell recognition of these integral membrane proteins is the same as that of extrinsic antigen has aroused much interest (*Figure 4.6*). It is possible that one domain of an MHC gene product is able to act as a restriction element presenting an adjacent region that engages the TCR. Since functional experiments suggest that alloreactive T cells recognize the polymorphic determinants of MHC molecules, such a model would suggest that either the conserved regions are restriction elements or that polymorphic determinants function as both antigen and restriction

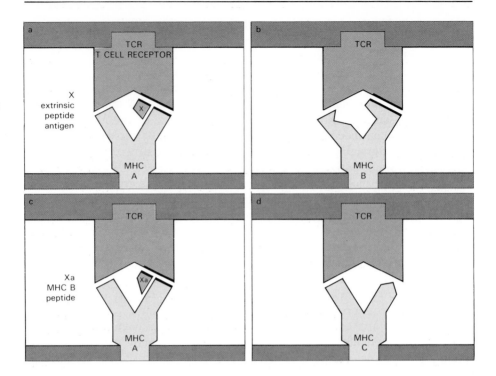

Figure 4.6. Models of T cell allo-antigen recognition. (**a**) Recognition of nominal antigen where peptide X binds to MHC of haplotype A and engages the α and β chains of the TCR. (**b**) Allo-antigen (MHC B) mimics MHC A plus X in conformation and so engages the TCR. (**c**) Allo-antigen peptide (Xa) binds to MHC A mimicking MHC A plus X, and (**d**) MHC C fails to stimulate the T cells.

elements (*Figure 4.6a* and *b*). Using the combined techniques of exon shuffling and gene transfection it has been demonstrated that MHC class I allospecific Tc recognize a combinatorial epitope formed by the α1 and α2 domains which also restricts viral antigen presentation (35). The same technical approach has identified the amino-terminal half of the β1 domain of MHC class II as the site of class II specific alloreactive T cells (36).

An alternative proposal is that MHC-encoded molecules are 'processed' and presented by accessory cells employing the same mechanisms as extrinsic antigen. To resolve this it is necessary to demonstrate that allorecognition is accessory-cell-dependent and that peptide fragments of allo-antigen in the presence of the appropriate restriction element can stimulate alloreactive T cells. Murine-class-I-restricted Tc clones (H-2^d) have been isolated that recognize human class I antigen (HLA-CW3) transfected into murine cells (12). These T cell clones lysed syngeneic spleen cells only in the presence of the relevant synthetic peptide, CW3 (residues 171–186). Unrelated peptide sequences of CW3 or MHC class I histo-incompatible (non-H-2K^d) target cells were not lysed. It has also been reported that peptides (residues 98–113) derived from the α2 (hypervariable) region of HLA-A2 can specifically inhibit the cytolytic activity of HLA-A2-specific, human T cell clones (37; *Figure 4.6c*). Thus, allospecific and viral-specific class-I-restricted Tc are similar in that both appear able to recognize peptide fragments of antigen and suggest that T cell recognition of extrinsic and allo-antigen is by the same mechanism. This has yet to be described for MHC class II allorecognition.

Since unrelated antigens can compete for antigen presentation, Guillet *et al.* (27) propose that the hypervariable regions of MHC class II molecules contain both an internal receptor and ligand and that complementarity exists between each chain of class II. For foreign peptides to bind to MHC class II they must displace the internal ligand and are thus seen by the TCR as analogues of self. To account for allorecognition this model requires that T cells recognize the internal ligand of foreign MHC class II molecules. Provided that the internal ligand contains polymorphic residues, then extrinsic antigen complexed to self and foreign MHC class II bound with its own internal ligand could not be distinguished by the TCR. Therefore, areas of homology or common motifs between the two ligands would induce an alloresponse. So allo-antigen appears the same as self plus extrinsic antigen. Evidence in support of this comes from the investigation of MYO-reactive, I-A^d-restricted T cells that also proliferate to I-A^b (38). Sequence homology exists between MYO and I-A^b at five residues in the hypervariable region (residues 1–20) and identity (H-12) may be responsible for I-A^b (allo-antigen) mimicking MYO/I-A^d in terms of T cell recognition. Thus T cell allorecognition may depend upon structural homology between ligands.

5. Non-responder status: holes in the repertoire versus lack of MHC binding

The ability of an animal to generate an immune response to a given antigen is regulated by the MHC, such that certain peptides in combination with a particular MHC gene product are non-immunogenic. The regulation of this non-responsiveness by immune response (Ir) genes is thought to operate through one of two distinct mechanisms. The defect may lie in the MHC class II molecule failing to bind the antigen, thus preventing activation of T cells with receptors of the appropriate specificity (determinant selection theory). Alternatively, they influence the development of the T cell repertoire in that many potential receptor combinations are not expanded during positive selection of recognition of self-MHC-encoded molecules. The net effect of this would be to limit the T cell repertoire so that certain combinations of antigen–MHC T cell clones with the appropriate TCR do not exist and therefore no immune response is observed.

Evidence to support determinant selection initially came from functional studies and is best illustrated by the results obtained on the response to cytochrome *c* (23). It was observed that T cells from B10.A(5R), a low-responder strain, are able to respond to p.Cytc peptide 81–104 in the presence of high-responder

Table 4.2. Binding of peptides to MHC class II molecules of H-2^d and H-2^k haplotype

Labelled peptide	Percentage of labelled peptide bound to:			
	A^d	E^d	A^k	E^k
OVA (323–339)	<u>11.8^a</u>	0.1	0.2	0.1
λ repr. (12–26)	<u>1.6</u>	8.9^a	0.3	<u>2.3</u>
HEL (46–61)	0.0	0.0	<u>35.2^a</u>	0.5
HEL (74–86)	2.0	2.3	<u>2.9</u>	1.7
HEL (81–96)	0.4	0.2	0.7	<u>1.1^a</u>
MYO (132–153)	0.8	<u>6.3^a</u>	0.5	0.7
p.Cytc (88–104)	0.6	1.2	1.7	<u>8.7^a</u>
m.Cytc (88–103)	0.1	1.0	1.7	<u>5.3^a</u>

^{a}Significant binding. Underlined figures indicate known MHC restrictions of those peptides. Based on data of Buus and colleagues (40).

B10.A-presenting cells. This implies that the defect lies in the MHC class II molecules of the B10.A(5R). Again, taking advantage of the response of the B10.A to p.Cytc (residues 43–58), peptide analogues were synthesized with substitutions at only those positions which interacted with the TCR and not the MHC. As all the analogues were immunogenic this was interpreted as evidence of a lack of holes in the T cell repertoire (39). Perhaps the most compelling evidence is derived from the binding of antigens to purified MHC class II molecules and competition of related and unrelated antigens for presentation. In the majority of instances the capacity of a particular class II allele to bind peptide antigen paralleled the responder status of that haplotype (*Table 4.2*). Additionally, peptides that inhibited the binding of another peptide to MHC class II were also effective in competing with accessory cell presentation of that peptide (40). The ability to distinguish self from non-self is a function of T cells and not of MHC gene products themselves. Therefore, T cells must be able to avoid potentially autoreactive MHC class II – antigen combinations and this could be achieved by clonal deletion. Interestingly, it was observed that the λ cI peptide (residues 12–26) was able to bind to both I-A^d and I-E^d, but is not restricted by I-E^d in the H-2^d haplotype (27). Although the λ repr. peptide contains the common motif, in contrast with I-E^d-restricted peptides, three residues are identical with the hypervariable β1 domain of I-E^d. As a result of these shared residues T cells capable of recognizing the λ repr. peptide may have been deleted during the induction of tolerance to self (I-E^d), which would explain the lack of immunogenicity of the λ repr. peptide – I-E^d complex. Consistent with this interpretation, that holes in the T cell repertoire exist, are the experiments that have identified T cell clones which are activated by antigen in association with MHC class II molecules from low-responder haplotypes (41,42). These experiments can be criticized from the point of view that the individual T cell clones analysed were unique and as such not representative of the T cell population as a whole. This was resolved by Ishii *et al.* (43) using populations of allogeneic T cells acutely depleted of alloreactive cells by treatment with bromodeoxyuridine and light. They observed that T cells from responder and non-responder strains were able to proliferate to antigen, the glutamic acid – alanine co-polymer (GA) in association with MHC class II of allogeneic non-responder macrophages in all strain combinations studied. Only T cells from a non-responder haplotype failed to respond to antigen in the presence of syngeneic or F1 hybrids with the same non-responder strain as one of the parents. These results were interpreted as evidence against non-responder status being a defect of MHC class II – antigen complex formation but an absence from the repertoire of T cells expressing the appropriate TCR.

In conclusion, there is evidence to support the two mechanisms of Ir gene regulation and, in reality, although both may operate, quantitatively determinant selection may be the most important.

6. Further reading

Berzofsky,J.A. (1987) In Sela,M. (ed.), *The Antigens*. Academic Press, New York.
Moller,G. (ed.) (1987) *Immunol. Rev.,* **98**.
Schwartz,R.H. (1986) *Adv. Immunol.,* **38**, 31.
Sercarz,E.E. and Berzofsky,J.A. (1987) *Immunogenicity of Protein Antigens: Repertoire and Regulation*. CRC Press, Boca Raton, FL.

7. References

1. Unanue,E.R. (1984) *Annu. Rev. Immunol.,* **2**, 395.
2. Ellner,J.J., Lipsky,P.E. and Rosenthal,A.S. (1977) *J. Immunol.,* **118**, 2053.
3. Gell,P.G.H. and Benacerraf,B. (1959) *Immunology,* **2**, 64.
4. Thomas,D.W., Hseieh,K.-H., Schauster,J.L. and Wilner,G.D. (1981) *J. Exp. Med.,* **153**, 583.
5. Buus,S., Sette,A. and Grey,H.M. (1987) *Immunol. Rev.,* **98**, 115.
6. Shimonkevitz,R., Kappler,J.W., Marrack,P. and Grey,H.M. (1983) *J. Exp. Med.,* **158**, 303.
7. Streicher,H.Z., Berkower,I.J., Busch,M., Gurd,F.R.N. and Berzofsky,J.A. (1984) *Proc. Natl. Acad. Sci. USA,* **81**, 6831.
8. Allen,P.M. (1987) *Immunol. Today,* **8**, 270.
9. Townsend,A.R.M., Gotch,F.M. and Davey,J. (1986) *Cell,* **42**, 457.
10. Townsend,A.R.M., Rothbard,J., Gotch,F.M., Bahadur,G., Wraith,D. and McMichael,A.J. (1986) *Cell,* **44**, 959.
11. McMichael,A.J., Gotch,F.M. and Rothbard,J. (1986) *J. Exp. Med.,* **164**, 1397.
12. Maryanski,J.L., Pala,P., Corradin,G., Jordan,B.R. and Cerottini,J.-C. (1986) *Nature,* **324**, 578.
13. DeLisi,C. and Berzofsky,J.A. (1985) *Proc. Natl. Acad. Sci. USA,* **82**, 7048.
14. Rothbard,J.B. (1986) *Ann. Inst. Pasteur,* **137E**, 497.
15. Watts,T., Brian,A., Kappler,J., Marrack,P. and McConnell,H. (1984) *Proc. Natl. Acad. Sci. USA,* **81**, 7564.
16. Werdlin,O. (1982) *J. Immunol.,* **129**, 1883.
17. Rock,K.L. and Benacerraf,B. (1983) *J. Exp. Med.,* **157**, 1618.
18. Erb,P. and Feldmann,M. (1975) *Eur. J. Immunol.,* **5**, 759.
19. Babbitt,B., Allen,P., Matsueda,G., Haber,E. and Unanue,E. (1985) *Nature,* **317**, 359.

20. Buus,S., Sette,A., Colon,S.M., Jenis,D.M. and Grey,H.M. (1986) *Cell,* **47**, 1071.
21. Sette,A., Buus,S., Colon,S., Smith,J.A., Miles,C. and Grey,H.M. (1987) *Nature,* **328**, 395.
22. Watts,T.H., Gaub,H.E. and McConnell,H.M. (1986) *Nature,* **320**, 179.
23. Schwartz,R.H. (1985) *Annu. Rev. Immunol.,* **3**, 237.
24. Allen,P.M., Matsueda,G.R., Evans,R.J., Dunbar,J.B., Marshall,G.R. and Unanue,E.R. (1987) *Nature,* **327**, 713.
25. Heber-Katz,E., Hansburg,D. and Schwartz,R.H. (1983) *J. Mol. Cell. Immunol.,* **1**, 3.
26. Guillet,J.-G., Lai,M.-Z., Briner,T.J., Smith,J.A. and Gefter,M.L. (1986) *Nature,* **324**, 260.
27. Guillet,J.-G., Lai,M.-Z., Briner,T.J., Buus,S., Sette,A., Grey,H.M., Smith,J.A. and Gefter,M.L. (1987) *Science,* **235**, 865.
28. Hansburg,D., Heber-Katz,E., Fairwell,T. and Appella,E. (1983) *J. Exp. Med.,* **158**, 25.
29. Ronchese,F., Brown,A.M. and Germain,R.N. (1987) *J. Immunol.,* **139**, 629.
30. Braunstein,N.S. and Germain,R.N. (1987) *Proc. Natl. Acad. Sci. USA,* **84**, 2921.
31. Kappler,J.W., Wade,T., White,J., Kushnir,E., Blackman,M., Bill,J., Roehm,N. and Marrack,P. (1987) *Cell,* **49**, 263.
32. Davis,M.M. (1985) *Annu. Rev. Immunol.,* **3**, 537.
33. Bjorkman,P.J., Saper,M.A., Samraoui,B., Bennett,W.S., Strominger,J.L. and Wiley,D.C. (1987) *Nature,* **329**, 506.
34. Bjorkman,P.J., Saper,M.A., Samaroui,B., Bennett,W.S., Strominger,J.L. and Wiley,D.C. (1987) *Nature,* **329**, 512.
35. Allen,H., Wraith,D., Pala,P., Askonas,B. and Flavell,R.A. (1984) *Nature,* **309**, 279.
36. Germain,R.N., Ashwell,J.D., Lechler,R.I., Margulies,D.H., Nickerson,D.H., Suzuki,G. and Tou,J.Y.L. (1985) *Proc. Natl. Acad. Sci. USA,* **82**, 2940.
37. Parham,P., Clayberger,C., Zorn,S.L., Ludwig,D.S., Schoolnick,G.K. and Krensky, A.M. (1987) *Nature,* **325**, 625.
38. Cease,K.B., Berkower,I., York-Jolley,J. and Berzofsky,J.A. (1986) *J. Exp. Med.,* **164**, 1779.
39. Ogasawara,K., Maloy,W.L. and Schwartz,R.H. (1987) *Nature,* **325**, 450.
40. Buus,S., Sette,A., Colon,S., Miles,C. and Grey,H.M. (1987) *Science,* **235**, 1353.
41. Clark,R.B. and Shevach,E.M. (1982) *J. Exp. Med.,* **155**, 635.
42. Kimoto,M., Krenz,T.J. and Fathman,C.G. (1981) *J. Exp. Med.,* **154**, 883.
43. Ishii,N., Baxevanis,C.N., Nagy,Z.A. and Klein,J. (1981) *J. Exp. Med.,* **154**, 978.

Glossary

Affinity: a measure of the strength of interaction between a single antigen receptor and its epitope.

Agretope: the part of an antigen which interacts with a MHC molecule.

Allo-antigen: an antigen recognized by lymphocytes in animals of the same species with different haplotypes.

Antigen: a molecule recognized by the immune system, which elicits an immune reaction.

Antigen processing and presentation: the function of a group of cells (antigen-presenting cells) which converts native antigen into a form in which it becomes associated with MHC molecules and can be recognized by T lymphocytes.

Antigen receptors: the molecules on lymphocytes which are responsible for specifically binding to and recognizing antigens or antigen/MHC.

Autoreactive: describes lymphocytes which recognize the individual's own molecules and mount an immune reaction to them.

C domain: the domains of antibody and the T cell receptor which show relatively little variability between molecules with different specificity and which do not contribute to antigen binding.

CD molecules: a system of nomenclature for lymphocyte surface molecules, including:
    CD2—present on T cells and involved in lymphocyte adhesion and antigen non-specific cell activation;
    CD3—present on T cells associated with the antigen receptor and involved in antigen specific cell activation;
    CD4—present primarily on helper T cells and involved in class-II-restricted interactions;
    CD5—present on some T cells, including T helpers, and a small subset of B cells (previously Ly1).
    CD8—present primarily on cytotoxic T cells and involved in class-I-restricted interactions;
    CD25—present on antigen-activated T and B cells acting as a receptor for IL-2.

Class I, class II molecules: two types of molecule encoded within the MHC involved in T-cell-mediated immune recognition.

Class-I/class-II-restriction: describes interactions in which class I or class II MHC molecules are involved (see MHC-restriction).

D gene: small genetic elements which are recombined with V and J genes to produce a gene for the V domains of antigen receptors.

Desetope: the part of an MHC molecule which binds to processed antigen.

Domain: a globular region of folded polypeptide. Antigen receptors and MHC molecules are formed into domains.

Epitope: the region of an antigen which binds to an antigen receptor.

Fab: a single arm of an antibody containing one antigen binding site.

Genetic restriction: see MHC-restriction.

H-2: The mouse MHC. Strains of different haplotype are designated with small superscripts, e.g. $H-2^k$, $H-2^d$ etc. H-2K and H-2D encode class I molecules, I-A and I-E encode class II molecules.

Haplotype: a set of genetic determinants treated on a single chromosome.

HLA: human MHC. A, B and C encode class I molecules while the D region encodes class II molecules.

Hypervariable region: small loops of polypeptide within the V domains of antibody which are most variable between different antibodies and which lie at the antigen binding site.

Ia: Class II molecules encoded by the I-A locus.

I-A/I-E: Class II gene loci of the mouse MHC.

Interleukins: a group of molecules involved in signalling between lymphocytes, antigen-presenting cells and other cells in the body.

J gene: see D gene.

LFA molecules: a group of molecules involved in antigen non-specific binding of lymphocytes to other cells.

MHC: the major histocompatibility complex is a gene locus encoding (among others) cell surface molecules involved in antigen-specific interactions between T lymphocytes and other cells.

MHC-restriction: the observation that cells must be of the same MHC haplotype to cooperate effectively. They are said to be genetically restricted. In some cases the requirement is that the cells share haplotype of class I molecules, and in other cases that the class II molecules are identical. The requirement for MHC-restriction implies that MHC molecules must be involved in the interaction.

Paratope: the part of an antigen receptor which specifically binds to antigen.

Pseudogenes: genes which are homologous in sequence to other expressed genes, but which are defective in some way, so that they themselves cannot be expressed.

Repertoire: the total range of antigen receptor specificities within an individual.

Somatic mutation: a process by which antibodies can refine the specificity of their binding by activating a mechanism which mutates antibody gene DNA.

Somatic recombination: a process of breaking and rejoining of DNA which is involved in the generation of genes for V domains.

Tolerance: the failure of lymphocytes to react to an antigen which they have recognized.

V domain: the N-terminal domains of the antibody and T cell receptor which are highly variable between different molecules and which form the antigen binding site.

V gene: see D gene.

Index